Paths

András Gulyás • Zalán Heszberger • József Biró

Paths

Why is life filled with so many detours?

 Birkhäuser

András Gulyás
Budapest University of Technology
and Economics
Budapest, Hungary

Zalán Heszberger
Budapest University of Technology
and Economics
Budapest, Hungary

József Biró
Budapest University of Technology
and Economics
Budapest, Hungary

ISBN 978-3-030-47547-5   ISBN 978-3-030-47545-1   (eBook)
https://doi.org/10.1007/978-3-030-47545-1

This book is an open access publication.

Cover illustration: Cover Image based on the original drawing by Lajosné F. Kiss and printed with his permission.

This book is published under the imprint Birkhäuser, www.birkhauser-science.com, by the registered company Springer Nature Switzerland AG.
The registered company address is: Gewerbestrasse 11, 6330 Cham, Switzerland

The path is the goal...—Mahatma Gandhi

To our loving wives and children:
Nusi, Bandika, Gabi
Andris, Dóri, Tündi
Lili and Zsófi

Foreword: Paths We Live By

During the last generation and with the advent of the interconnectedness of people, institutions, and ideas through computer webs, network-based metatheories of all aspects of sciences started to flourish from metabolic mechanisms, to authors co-citations. Now, network science of different levels of abstraction flourishes in mathematics, physical, and biological models, as well as in sociological theories. There are many shining stars of Hungarian science on this path, from László Lovász, László Barabási Albert, János Kertész to Péter Csermely and György Buzsáki. This little book of another Hungarian trio—András Gulyás, Zalán Heszberger, and József Bíró—takes another look at these developments. Their perspective is not the network itself but the routes taken by neural firing patterns, handshakes, or word activations to arrive from one node in a network to another.

Paths have been the central idea of many social sciences for more than a hundred years. One of the most fundamental methods of the comparative psychology of animal cognition has been and continues to be the maze learning introduced in 1901 to psychology. For a long time, we treated it as a way to study the universal mechanism of learning. Today, we realize that it is the key to understand how mammals are able to internalize a map of different possible paths in their world full of orientational cues and object valences. Paths are used by animals to arrange the knowledge about their activities like where to go and what to do.

These research paths led to the search for neural paths in the brain by assuming specialized brain structures responsible for the long time assumed cognitive maps. This book presents the paths connecting the words in the lexicons and in the mind, the paths leading from corona to death during the pandemic. Several researchers, including György Buzsáki at New York University, hope that these later cognitive, meaning-based paths are tied to the same brain networks as the forest paths of the animals.

This book also presents the third important path system in humans, the one that takes one through the common past: instrumental and emotional contacts to another person.

This readable and easily accessible little book fills the reader with hopes and promises towards the future of network research where paths shall be found to relate the personal, conceptual, and neural networks.

May 3, 2020, at Budakeszi, Hungary. In the middle of the coronavirus lockdown

Psychologist and Linguist Member of the   Csaba Pléh
Hungarian Academy of Sciences and Academia Europaea
Budakeszi, Hungary

Foreword: The Longest Journey

"Tell me, Master, is there a single word that one can follow throughout one's life?"—turned once one of his disciples to *Confucius*. The master replied, *"Isn't mutuality such a word?"* The disciple bowed silently and left contentedly.

The great and solemn word of mutuality is also known in our Western culture; most often, it denotes some bilateral relationship. This relationship is mostly considered valid by the partners for the duration of a specific ongoing action. With its announced application, they demonstrate that they take each other's aspects into account to the maximum. Mutuality is mainly used in connections between *you* and *me*, or *me* and the *others*. (According to many, this lean bilateralism is rooted deep in the idea of monotheism.) After the action (discourse or act) between the two parties takes place, the concept becomes invalid and practically ceases until the next situation.

In the interpretation of Confucius and his followers, mutuality is a much more meaningful word. In Chinese antiquity, this term referred to an entire network of *mutual relationships*, a combination of paths and detours, decisions, and choices, along with the consequences and repercussions that follow them. In contrast to the Western-style action-like, casual-use interpretation, Confucius and his followers never treated mutuality as a restricted bilateral relationship. For them, mutuality was a deep principle underlying the Universe, a *Weltanschauung*.

This ancient Asian approach has another defining feature: The *correlative approach* with *modal logical judgment*. Such thinking focuses on the correlated relationship of *adjacent* things and allows for *multiple* valid judgments (statements) at the same time. This is in high contrast to the *Western binary logic*, where only one of the two statements can be correct, rejecting the other as incorrect. Any third statement (if raised at all) is ruled out. This is the principle of the excluded third by Aristotle. In any decision-making situation, the ancient Asian logic, however, always recommends the application of the *intermediate third's* law. This law means that three or even more statements can be valid simultaneously. Such prudence of minding all chances comes from the way of seeing things to be interdependent (or in correlation to each other).

Well, this is precisely the principle that has been missing from European thinking since time immemorial! It takes into account the intermediate third and fourth statements: the permissive idea of the concurrent validity of more than one statement. In our decision-making situations, since *Aristotle*, *we* consider a statement to be valid (correct) or not *with no further option*.

I wonder why? Because we need the most effective solution in all cases. Decisions that do not provide the most effective and quickest solution are considered to be detours or misguided paths in the eyes of the progress-hungry, impatient Western hero.

This way of thinking took us to where we are. We have progressed, progressed undoubtedly, but maybe too fast, so fast that we have probably run over the finish line already. And there is no way back; it is impossible to correct. Our only option is to slow down, that is, if we have a drop of wit, at least we do not rush into a not at all promising future robbed by increased efficiency, effectiveness, growth, and almost completely deprived of our physical and mental living conditions.

We do it smarter if we slow down our progress, if we choose a detour to our goals. This slowdown gives us a chance to keep a common sense of the concept of mutuality in the broadest sense, an opportunity to be attentive, smart, and even to spare ourselves and each other (not just action-like). During the delicate, careful trying, and enjoyable tasting of the paths and detours, we may even get to know human nature better. So far, we did not have time to figure out whether such a thing exists at all. *"Human science would finally be needed,"* sighed *Ortega y Gasset* bitterly after *the* second great suicide attempt of the *genus humanum in* the twentieth century.

And if we understand the poet's word better, let us listen to *Constantine Cavafy*, who advises that *"When you set out on your journey to Ithaca, pray that the road is long, full of adventure, full of knowledge."* The longest journey gives you the greatest gift of the city.

This book is the wisest guide to paths: an easy-to-understand book of intact, unharmed presence, orientation, and amicable arrangement in the risky world of a million choices. It provides exactly the prudent and relational-centered approach that Western thinking needs the most today.

Orientalist, Writer, Hungarian Media Person   László Sári (a.k.a. Su-la-ce)
Budapest, Hungary
April 10, 2020

Acknowledgements

We would like to say special thanks to our parents and grandparents for showing us their paths and helping to smooth ours. We must express our very profound gratitude for many of the useful ideas and for the fruitful discussions to Zsófia Varga, Attila Csoma, Antal Heszberger, Gabriella F. Kiss, Attila Kőrösi, István Pelle, Dávid Szabó, and Gábor Rétvári. We are grateful for the careful perusal and kind comments of István Papp, Csaba Hős, Claudia Molnár, Attila Mertzell, László Gulyás, Mariann Slíz, Alessandra Griffa, Andrea Avena-Königsberger, Levente Csikor, Márton Novák, Dávid Klajbár, Valentina Halasi, Máté Csigi, Erzsébet Győri, Tamás Csikány, István Bartolits, Alija Pasic, Alexandra Balogh, Rudolf Horváth, and Mária Marczinkó.

A special thanks go to Anne Comment, our ever-patient Publishing Manager in coordinating the book project, and to Kathleen Moriarty, our Copy Editor for her careful reading and invaluable help in revising the final manuscript. Special gratitude to Lajosné F. Kiss for creating the fantastic illustrations.

Disclaimer

This book does not intend to communicate any scientific consensus about paths. In fact, there is no consensus about paths. Ideas presented here, although mostly founded on real-world data, primarily reflect the authors' subjective (sometimes speculative) image about the world. This work is intended to entertain, inspire, and persuade the reader to think critically about the nature of paths as taken by people as well as many other entities in life.

Contents

Content Path

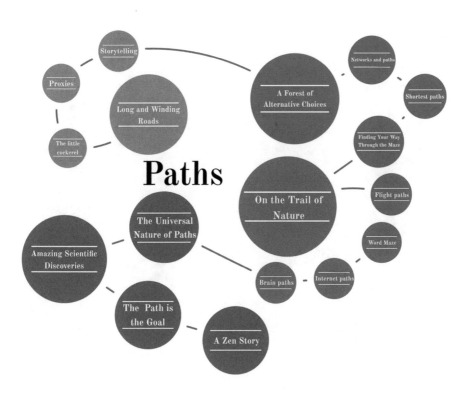

Storytelling

Proxies

Long and Winding Roads

The little cockerel

A Forest of Alternative Choices

Networks and paths

Shortest paths

Finding Your Way Through the Maze

Flight paths

Paths

On the Trail of Nature

The Universal Nature of Paths

Word Maze

Amazing Scientific Discoveries

Brain paths

Internet paths

The Path is the Goal

A Zen Story

List of Figures

List of Tables

Chapter 1
Introduction: Long and Winding Roads

Once upon a time, *there was a cock and a mouse. One day the mouse said to the cock, "Friend cock, shall we go and eat some nuts on yonder tree?" "As you like." So they both went under the tree, and the mouse climbed up at once and began to eat. The poor cock began to fly, and flew and flew, but could not come where the mouse was. When it saw that there was no hope of getting there, it said, "Friend mouse, do you know what I want you to do? Throw me a nut." The mouse went and threw one and hit the cock on the head. The poor cock, with its head broken and all covered with blood, went away to an old woman. "Old aunt, give me some rags to cure my head." "If you will give me two hairs, I will give you the rags." The cock went away to a dog. "Dog, give me some hairs. The hairs I will give the old woman. The old woman will give me rags to cure my head." "If you will give me a little bread," said the dog, "I will give you the hairs." The cock went away to a baker. "Baker, give me bread. I will give the bread to the dog. The dog will give hairs. The hairs I will carry to the old woman. The old woman will give me rags to cure my head. [5] . . . "* (Fig. 1.1).

We could go on with the story, but to quickly reassure the reader we state that the poor cock finally managed to cure his head after going through several other interesting adventures in the forest. Telling such cumulative tales to children is always great fun. They quickly pick up the rhythm of the story and listen to you with curious eyes throughout. But what makes those cumulative tales, like the Italian one above, so fascinating that children always listen and watch intently? Well, of course they are worried about the little cockerel and wonder if he can cure his head. But if that is all, then the tale could end after the nut hit the cock on the head by saying that "The poor cock, with its head broken and all covered with blood, went away to an old woman who gave him rags, and the cock cured his head." Not so brilliant. If we put it this way, the story would lose its meaning–its essence. But, what is at the heart of the tale that makes it exciting? We could say, a long chain of events that has to happen before the cock can finally heal his head. An intricate *path* of events which can take unexpected turns and may go on forever. A path which we go down with the little cockerel and almost forget why he desperately needs all of

© The Author(s) 2021
A. Gulyás et al., *Paths*, https://doi.org/10.1007/978-3-030-47545-1_1

Fig. 1.1 The cover of Italian popular tales by Thomas Frederic Crane [Published by the Riverside Press, Cambridge, Massachusetts, 1885]

ITALIAN POPULAR TALES

BY

THOMAS FREDERICK CRANE, A. M.
PROFESSOR OF THE ROMANCE LANGUAGES
IN CORNELL UNIVERSITY

BOSTON AND NEW YORK
HOUGHTON, MIFFLIN AND COMPANY
The Riverside Press, Cambridge
1885

those things. When listening to the tale, we are so preoccupied with following his path, that the goal almost vanishes from our horizon.[1] The whole adventure slowly becomes to exist in its own right, perhaps more important than the goal itself, and gains its own, independent meaning. Does it mean that we avoid getting to the goal? Well, not exactly. Wandering around pointlessly would become tiresome over time. But we seem to have a strange desire to meander a bit before finishing the story. Is it maybe to warm up or to attune ourselves to the story? Or is it simply a quest for some pleasure? Or do we just need time to prepare for an important message? Regardless of the reasons, the path eventually becomes the essence of the story, and the goal loses its meaning entirely!

If you have ever watched the classic Columbo crime series with Peter Falk, you will surely understand this idea. Each episode of Columbo starts by showing a murder exactly as it happened. So, from the very beginning, we know who the victim is, who the murderer is and how the murder has been committed. The ending of the story is not a question: Columbo will arrest the murderer. So, we don't watch this series for the excitement of whether the murderer will be caught or not. Then why do we watch it? Well, for the specific way Columbo solves the crime with all the tiny, seemingly insignificant details that are slowly pieced together to create an unwavering proof. In short, we watch it for Columbo's particular *path* towards solving the case. And, of course, for one more thing: Columbo's rigorously funny character.

---

[1]This thought is beautifully captured by the painting of Ma Yuan, where the figure in the painting walks on a mountain path, quickly vanishes (see Fig. 1.2).

Chapter 2
Everybody Loves Roundabouts

On a typical day, before starting an activity, we set an explicit or implicit goal. The path that leads towards that goal is often selected without carefully designing it. It just comes naturally to select the right route to work or the appropriate series of actions to prepare breakfast. It is no wonder that an optimal execution is rarely considered. There are, however, cases when our target is in some distant future, giving us an opportunity to mull over it and to discover an energy-saving solution. It comes as a surprise that we rarely take advantage of it. To us, the adopted path seems to contain superfluous steps requiring extra effort. What is more, sometimes people cannot stand making things more complicated. Is that simple human negligence or is there more to be discovered? In the following, let us make an attempt to unfold the mystery through a series of real-world examples. We first start with a story from the postwar America, and then we study human activities on the Internet, and finally we discuss a superb idea for project presentations.

2.1 Hiding Behind Proxies

1947–1960 was not an easy period for Hollywood artists, writers, and directors. After the beginning of the cold war, the political witch-hunt in search of communists culminated in the primitive act of blacklisting more than 300 artists as they were accused of having communist ties or sympathies. Orson Welles, Arthur Miller, Charlie Chaplin are just a few names who had lost their jobs and reputations because they were blacklisted. To continue their careers, many of the blacklisted wrote under the names of friends who posed as the actual writers. These friends were called "fronts". The motion picture titled "The Front" (1976), directed by Martin Ritt, was based on these regrettable events. Howards Prince (Woody Allen) (Fig. 2.1), the restaurant cashier and illegal bookie, is asked by his friend Alfred Miller (Michael Murphy), the blacklisted screenwriter, to sign his name to Miller's television scripts. Howard is a good friend and desperate for money and success, so he agrees. Miller's

© The Author(s) 2021
A. Gulyás et al., *Paths*, https://doi.org/10.1007/978-3-030-47545-1_2

Fig. 2.1 Howard Prince's (played by Woody Allen) portrait hand-drawn by Lajosné F. Kiss [With the permission of Lajosné F. Kiss]

scripts make both of them wealthy and Howard becomes a famous screenwriter. Howard proves to be a good "front" for Miller and attracts other "clients". He proceeds to publish their scripts under his own name. The business is booming and Howard becomes one of the most prominent screenwriters in Hollywood.

What a surreal way of becoming successful and famous! If we start to think a bit deeper about Howard's success, it becomes even more surreal. Why do the writers turn to him with their problems? Restaurant cashiers are not the typical supporters of writers in Hollywood. Directors, artists, businessmen are more likely to have the resources writers need. They have the money, social contacts, influence, and reputation. Howard does not have any of those. Despite that, many writers turn to Howard for help. What can Howard provide then, which is so valuable to Hollywood's famous writers? Instead of money, contacts or influence, he could have offered his harmless personality. He was a nobody and that is exactly what the writers needed. Through this representative, the desperate writers reached their goals: publish their scripts and continue their careers. Acting as a *front* for others, Howard simply provided a *path*; path to the goals, which otherwise would have been unreachable.

"Fronts", like Howard Prince, are widely used on the Internet too. They are called proxies, and in the most basic setting, they can be used to act as "straw men" for users when accessing an Internet service. In the example Fig. 2.2, Bob is running a current time service, meaning that he tells the current time for a few cents if someone asks. Alice uses a proxy to ask Bob the time. Why does she do it? First of all, to save some money. Alice has a friend named Carol who also uses Bob's service very frequently. They agree to use the same proxy to ask Bob the time. If Alice and Carol are curious about the time almost simultaneously, then the proxy can ask Bob once and tell both of the girls. The more people there are eager to know the time, the more money they can save by using the proxy. Secondly, as the proxy acts on behalf of Alice, Bob will never know that Alice uses his service; only knows the proxy.

Fig. 2.2 The working of a
proxy server

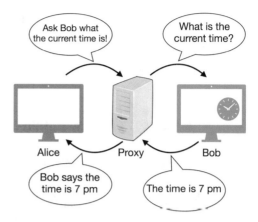

In 2003 Vivek Pai and his colleagues at Princeton University decided to set up many open proxies [22] all over the Internet for research purposes. Open means that the proxies can be used by anybody to indirectly access Internet services, like Alice and Carol did. Unfortunately open servers on the Internet are a hackers delight, so their intentional release was not a good idea. The researchers assumed that "an unpublicized, experimental research network of proxies would not be of much interest to anyone". They were wrong. They underestimated how long it would take for others to discover their system, and the scope of activities for which people sought open proxies. Vivek and his colleagues experienced extraordinary attention in their system immediately after its launch. They quickly detected a very large volume of traffic (emails, chats, downloads, casts) going through their open proxies.

The question is, can a non-advertised open proxy server farm draw the attention of anybody? We can agree with the researchers, that it is hard to imagine that such a seemingly valueless thing would interest more than a few moping networking fellows. What is it in this system that is so attractive to a surprisingly large amount of people? What does this system provide that is useful to many people with such diverse purposes?

Among the unforeseen activities of the open proxy experiment, the researchers observed that proxies in Washington and California received a very high amount of connections with both sources and destinations located along the eastern rim of Asia. The multi-megabyte downloads appeared to be for movies, though the reason that these clients chose round-trip access across the Pacific Ocean was not clear. A direct connection would presumably be much faster. A reasonable explanation for this behaviour is that these clients were banned from some websites and required fast proxies to access them without disclosing their identity. Given the high international Internet costs in Asia, proxies in the Western United States were probably easier to find. Regardless of the real motivation of these users, they all picked characteristic *paths* through the Internet to reach their services, their goals. In fact in this story, the paths play a more important role than the users and services they connect. The path, visible on an Internet map, in itself means something. It has its own reason for existence and tells us how people try to solve their problems, how they think and how they manage their lives. Besides financial or legal causes, there can be other more elusive human motivations to create perplexing systems of paths, like presenting complex ideas.

2.2   Mind Maps: The Revolution of Presenting Ideas

Showing a sequence of slides is the most widespread way of presenting ideas to an audience. In a slide-based presentation, the speaker goes through the slides, supporting the talk, in a linear fashion. Besides this mainstream slide-oriented approach, a new wave of storytelling tools have appeared in the market, centered around so-called mind maps. The users of such tools (e.g., Prezi, Mindmeister) can collect various materials (texts, images, videos, slides) related to a specific topic and organize them like a "mind map". A mind map is a drawing that visually organizes information. It is generally hierarchical and shows relationships among pieces of the whole (see Fig. 2.3). It is often created around a single concept, drawn in the center of the map, to which associated representations of ideas such as images, texts, videos and slides are added. Major ideas are connected directly to the central concept, and other ideas branch out from those.

The mind map oriented presentation approach quickly became popular among presenters and continues to attract millions of users. Mind maps are indeed beautiful and eye-catching, but it is hard to think that the main motivation of millions of users is to draw and present aesthetically appealing, didactic mind maps. Is it the pure concept of mind maps that enabled small startups to compete with giants of the IT industry like Microsoft, Google and Apple in the area of presentation softwares? Or is there more to it than meets the eye?

When you present using a mind map tool, you can define a presentation *path* through your mind map and only focus on parts of the whole map that are related to your specific talk. Your particular message depends on the audience. For example, explaining Newton's second law to elementary school students requires a fundamentally different path than presenting the same in a university lecture, although they can share common parts as well (e.g., illustrative figures, experiment descriptions). The identification of paths as the main tools of storytelling is one of the core innovations of mind map based approaches. Arguably, this feature is what draws the attention of millions of users. A story can be told in many ways, and each version holds a specific footprint, a particular path lying in the background. Are there good or bad paths in storytelling? What is the difference between them? That is the million dollar question to answer. One thing is certain: getting straight to the point is rarely didactic; making detours for instructive examples can be key to a successful presentation. Everybody loves roundabouts!

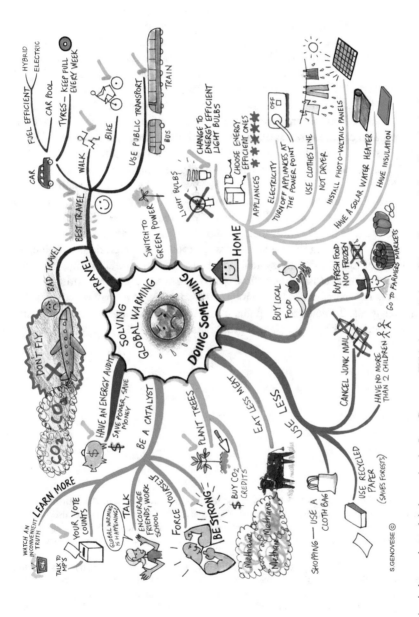

Fig. 2.3 A mind map about global warming by Jane Genovese [With the permission of Jane Genovese.]. http://learningfundamentals.com.au/

2.3 Short but Winding Roads

Despite their apparent independence, our nursery tale about the little cockerel, Howard Prince, the open proxy system and mind map based presentation tools share something which makes them so compelling that they attract much attention. It turns out that, although in different forms of appearance, they all provide different forms of *paths*. Paths to entertain us, paths to reach our goals, paths to solve our problems and paths to deliver our messages. These examples indicate that paths play an important role in diverse areas of life. It seems that paths are somewhat universal. They are abstractions which can emerge in various kinds of guises. Is it possible that the paths coming from different areas share some common properties? Is it possible that these paths are the product of some general laws that can be identified? What are the possible steps that a path might include? These are difficult questions to tackle. First, we can make an interesting observation. Common sense suggests that we should favor "short" paths. We don't like lengthy talks, we don't want to forget our goals when seeking a path, and we don't have infinite time and energy to solve our problems. Does it automatically mean that we should use the "shortest" possible path? Our earlier examples hint that the shortest path may not always be the best choice either. Shortening the nursery tale will diminish its entertainment value, the Asian users of the open proxy system do not use a shorter direct path and a short talk concentrating exclusively on the essentials of a topic may be boring or hard to interpret. They should not be too long, but some windings may be necessary to reach their goals. To analyze our paths in the following chapters, we need formal concepts to grasp their most essential properties.

Chapter 3
The Forest of Alternative Choices

Watch the path of your feet And all your ways will be established.

—Proverbs 4:26

In general, a path can be thought of as a sequence, timely ordered sequence of consecutive events or choices which can lead us far from the starting point.

Following a path means that we choose a specific sequence of steps from a pool of possibilities or alternative choices. So what does such a pool look like? Well, sometimes it is small, concrete and well-defined, while other times it is seemingly infinite and may be obscure and intricate. To illustrate, let's consider a very famous and classic event pool from European century.

In the eighteenth century, the city of Königsberg, Prussia was wealthy enough to have seven bridges across the river Pregel. The seven bridges connected four parts of lands separated by the branches of the river. The situation is shown in Fig. 3.1 where letters A, B, C, D denote the lands and the corresponding handwriting (ending with the B. and Br. abbreviations) mark the locations of the bridges. This scenario inspired the fantasy of the leisured inhabitants of Königsberg who made a virtual playground from the bridges and lands. Their favorite game was to think about a possible walk around the bridges and lands in which they cross over each bridge once and only once. Nobody could come up with such a fancy walk and nobody managed to prove that such a walk was impossible, until Leonhard Euler, the famous mathematician, took a look at the problem. Euler quickly noticed that from the perspective of the problem, most of the details of the map shown in Fig. 3.1 could be omitted and a much simpler figure could be drawn, focusing more on the problem (see Fig. 3.2).

This new representation contains only "nodes" marked with letters A, B, C, D in circles which represent the lands and "edges" drawn with curved lines between the nodes representing the bridges. A walk now can be described as a sequence of nodes and edges. For example the sequence $A \rightarrow E1 \rightarrow C \rightarrow E3 \rightarrow D \rightarrow E4 \rightarrow A$ represents a walk starting from land A which proceeds to land C via bridge E1, then to land D via bridge E3, and finally back to land A via bridge E4. Using only the nodes and edges, all sorts of walks can be created. In fact, all the possible walks

© The Author(s) 2021
A. Gulyás et al., *Paths*, https://doi.org/10.1007/978-3-030-47545-1_3

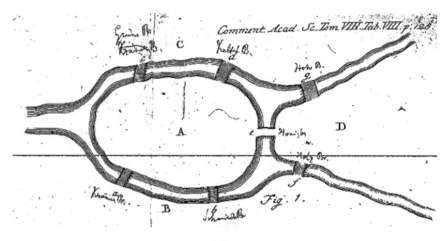

Fig. 3.1 Euler's Fig. 1 for the seven bridges of Königsberg problem from 'Solutio problematis ad geometriam situs pertinentis,' Eneström 53 [source: MAA Euler Archive; http://eulerarchive.maa. org/docs/originals/E053.pdf]

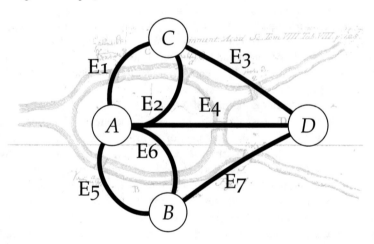

Fig. 3.2 Euler's idea of abstracting away the network underlying the Seven Bridges of Königsberg puzzle

that one can imagine throughout the bridges and lands is captured by this simple representation. The collection of nodes and edges called a network $N(n, e)$ turned out to be so powerful in modeling real-world problems that a whole new branch of mathematics, called graph theory,[1] was defined based on them.

---

[1]In the first ever graph theoretic argumentation Euler showed that to find a walk crossing each bridge once and only once requires that the underlying network can contain only two nodes with an odd number of edges. In Fig. 3.2 one can see that all nodes have an odd number of edges (A has five, while B, C and D has three), which makes the problem insolvable in this network.

We can observe that the walks around the lands and bridges are nothing more than the ordered sequences of consecutive events (bridge crossings) in Königsberg. These walks are very similar to our *paths* and the network $N(n, e)$ seems to effectively contain all the possible paths that can be taken, i.e., the paths[2] that can be differentiated by the sequence of bridge crossings in Königsberg. So, a network seems to be a good representation of the pools from which paths can take shape. The network in the case of the Königsberg bridges is very small and well-defined (contains four nodes and seven edges); however, the number of possible paths that people can take in this network is theoretically infinite as the length of the paths is not limited. In practice, the pool of all possible paths is much smaller as people become tired or bored after a few hours of walking. Even if we remove E2 and E3 and we are allowed to cross a maximum of ten bridges during the walk, there are still 2330 possible paths to choose from. Would people generally have a preference when choosing their afternoon walk? Will they choose randomly from all the possibilities? Or is there a hidden order affecting their choices? Those questions get even more complicated when, as in many real-life situations, a few more orders of magnitude of choices are at hand. For the sake of extending our scope for other connected systems, let us take a slightly more abstract network from the social sciences.

The small world experiments conducted by Stanley Milgram, a famous social psychologist, in the 1960s targeted to understand the network of human contacts in society. The main goal was to study the connectedness of people formed by their acquaintances. In the experiments, several random people were asked independently to send a letter (postcard) to a randomly chosen common target. Anyone, who did not personally know the target, was asked to send the letter to a friend who possibly would. Then the selected friends were subsequently asked to act the same way until the message arrived. In such a way, the persons were the nodes, and the friendships made up the edges of the network, while the traveler was the letter. Although many of the messages never arrived, those that did found a surprisingly short way through the chain of acquaintances. In many cases, even two or three middlemen were just enough for the letter to arrive (the average path length fell close to 6), in spite of the fact that the endpoints of the chain were carefully picked to be sufficiently far away from each other either geographically or socially. The so-called small world phenomenon is quite arresting in itself, the way the path is formed on the social graph is also fascinating. The arbiter behind the wanderer in this case is not a single entity but several independent ones, thus the established route is a collective phenomenon. Is there any similarity between the paths taken by individuals or the members of a community having a partially divergent perception on their environment? Do they achieve better at finding the shortest paths? Or do they require some superfluous sidesteps as well?

We will be able to answer such questions soon, but for now, let's be satisfied with finding networks as good representatives of all the imaginable paths belonging

---

[2]Note that the word path as used in this book corresponds to walks in the terminology of graph theory.

to a specific situation, because we will use them throughout this book. So, we have
networks over which one can take paths by traversing nodes and edges in a particular
sequence. But who or what will take the paths? Well, sometimes they are people as
in the Bridges of Königsberg problem, sometimes a letter controlled by a small
community as in the Milgram experiments. But in a broader scope, there can be
many things that can take paths. Gossip, fashion styles, memes and all sorts of
information seem to travel over social networks. If we look inside the human brain,
we can identify the neurons as nodes and their axons as edges. What travels through
this network? All kinds of information encoded into the specific firing patterns of
neural cells. Similarly to networks (which can represent all kinds of paths), we need
to find a name for the something which will travel through the network. From now
on, we will call these travelers "packets". Networks and packets will be all we need
to discuss paths in the broadest scope. Now let's consider a much more intricate
network, over which the traveling "packets" will be indeed: packets.

The Internet is the greatest network man has ever built. Starting from a small
research network funded by the US government, it became a huge interconnection
network of thousands of computers all over the world. In its early phase, the Internet
was similar in size to the network lying behind the bridges of Königsberg problem.
It had so few nodes and edges that one could draw its map on a single piece of paper
(see Fig. 3.3). After opening the network to the rest of the world, making it possible
for almost anybody to connect, an interesting game began which still goes on today.

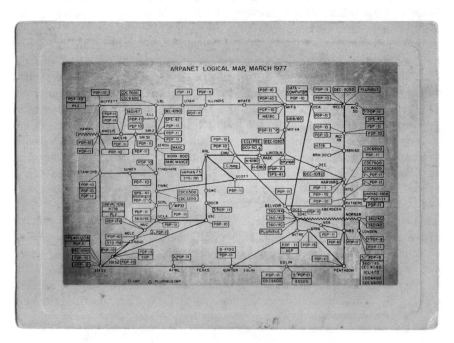

Fig. 3.3 Logical map of the ARPANET (the ancestor of the Internet) from 1977 [source: The
Computer History Museum; https://computerhistory.org/]

Nodes started to join the network in an uncoordinated fashion, which meant that nodes and edges could have appeared almost anywhere in the network. As a result of this process, the Internet evolved into a large, complex network, the topology of which changes heavily day-by-day. Even drawing an approximate contemporary map was a great challenge for networking researchers and its visualization needed newly developed algorithms. Despite the researchers' best efforts, such maps were able to grasp only a limited subset of the edges present on the Internet. Above this large and evolving network, our emails, chat texts, web pages and videos travel day-by-day. All these data, converted to small information packets, are delivered through paths determined by the Internet's so-called "routing" system. This routing system has no central authority which could compute the paths for every single packet. Quite the contrary, the Internet's routing system is heavily decentralized, meaning paths are determined through the complex interactions of thousands of nodes. What kinds of paths come out of such a process? We know that latency is crucial if it comes to Internet services. Nothing is more irritating than a website that is slow to respond, a lagging video conference or a frozen video game. So it is natural that we expect the provisioning of low latency paths. But does it mean that we will have the shortest path between our computer and the desired service? Recalling the example of the Asian users of the open proxy system can make us suspicious. As usual, the truth will lie between the two extremes. But what exactly are these shortest paths? Now it is time to get acquainted with them.

Chapter 4
Straight to the Point: A Short Chapter About the Shortest Paths

A passenger walks in the Őrség (region of Hungary) and asks a man mowing at the fringe of the forest in Szalafő: – How far is Őriszentpéter from here? – Five kilometers in a straight line, but I can get a shorter ride through the woods.

—*György Moldova, Tökös-mákos rétes, Magvető 1982*

There is something compelling about shortest paths. They are so simple and reasonable. They seem to be the most efficient paths for traveling between nodes in a network. They may take the lowest amount of distance, time or energy. For grasping the idea of shortest paths, let's consider the network in Fig. 4.1. In this network, the shortest path between nodes D and H is the path (D → C → E → G → H) marked with red arrows. Its length is the number of edges crossed which is 4 and this is the only shortest path between D and H. Green arrows mark the shortest paths from node C to node F. There are two shortest paths (C → B → A → F) and (C → E → G → F) and both have a length of 3. Shortest paths are also pretty straightforward to compute by a few lines of code e.g., by using Edsger W. Dijkstra's [7] method.

The compelling concept of shortest paths makes them first-class citizens in many areas of life. Everybody tries to take the shortest path from the store to the car or from home to the workplace, to save time and energy. Engineers of computer networks kindly favor the shortest paths because of their low latency and low resource usage (they load the smallest possible amount of routers and links). Shortest paths are also kindly used to predict information flow in social, biological and transportation networks. Researchers also use them to categorize networks and predict their behaviour under unusual circumstances (e.g., testing the behavior of the Internet during a massive natural disaster).

Although shortest paths are definitely desirable, there are also some problems with them. First, to find the shortest paths, one needs to explicitly know the whole network. Any program computing shortest paths requires the whole network as an input to run. To illustrate how much the shortest paths may change, imagine that we forgot a single edge in Fig. 4.1, namely the C → G edge, which is drawn with a dashed line in Fig. 4.2. In this modified network, the red path is not the shortest path between D and H anymore, since the path D → C → G → H is shorter. The shortest

A. Gulyás et al., *Paths*, https://doi.org/10.1007/978-3-030-47545-1_4

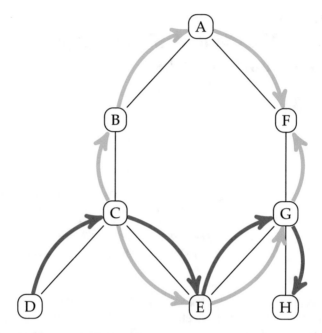

Fig. 4.1 Shortest paths on a simple network

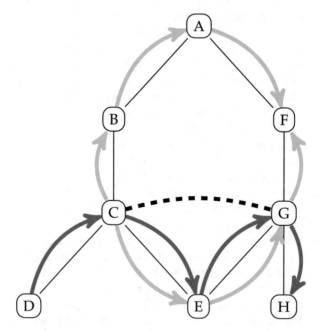

Fig. 4.2 The variability of shortest paths

path between C and F is neither of the green paths since $C \to G \to F$ is shorter than both of them. In addition, in this new situation there is only one shortest path between C and F.

Well, providing the exact structure of the network is not a problem in the case of small and quasi-static networks (e.g., the Bridges of Königsberg), but it is clearly not an option for large, complex and dynamic networks like the Internet. Secondly, there is something strange, something artificial in shortest paths. It seems that shortest paths sometimes fall too short and do not coincide with the underlying natural logic of the network (just think about the little cock or the users of the open proxy system or mind map presentations). Wait a minute! Can a network have a natural internal logic? Let's take a closer look at how such a logic might look!

Chapter 5
Finding Your Way Through the Maze

Orr was crazy and could be grounded. All he had to do was ask; and as soon as he did, he would no longer be crazy and would have to fly more missions. Orr would be crazy to fly more missions and sane if he didn't, but if he was sane, he would have to fly them. If he flew them, he was crazy and didn't have to; but if he didn't want to, he was sane and had to.

—Joseph Heller, Catch-22, Simon & Schuster 1961

Have you ever wondered how you would be able to navigate yourself through the labyrinthine street network of a town without any central knowledge base like a map or a GPS device? One thing is sure, to wander around would result in an inadmissibly long journey, even in a smaller settlement. How about the letter in the social acquaintance network of Milgram's small world experiments? Is it a feasible scenario that the letter just accidentally finds its way towards an addressee without any central guidance for the messengers passing it randomly to each other? With only a maximum of ten bridges across Königsberg's four islands resulted in 2330 different paths; what would happen in a network containing 300 million nodes with a hundred times more edges between them? The turmoil would be inconceivable! The very existence of short paths between the nodes of a network is one thing, to find them is completely another thing. It is reasonable to assume that there must be some landmarks or traffic signs in even the smaller networks if we are to find an adequately short way through it. There must be some internal logic that helps us to navigate from node to node towards our predefined destination without spending too much time roaming in the maze.

The internal logic of a network is something that is, on one hand, strongly connected to its outlook or construction. However there is sometimes something that is even more important than that: it is the rules of how to use paths among the nodes. In real networks, it is not uncommon that, although a path exists between nodes, it cannot be used due to some rules. For example think about a traffic sign that indicates one cannot enter a road unless invited by a resident. Just like the sign at the house of Winnie the Pooh's friend Piglet: "TRESPASSERS W." (Or was it really the name of his grandfather?) Or what about a carpool lane where a path can

© The Author(s) 2021
A. Gulyás et al., *Paths*, https://doi.org/10.1007/978-3-030-47545-1_5

only be used by cars shared by multiple travelers? Those also very much belong to the internal logic of a network: *it is about how a network may be used*. In the following let us take a look at some more complex examples, one from history and one from technology: the military organization network, and the Internet.

Military organizations have a strong internal logic: a *hierarchy*. As we will see, this strict hierarchy has a fundamental influence on the internal communication paths. The network representation of an imaginary military organization is shown in Fig. 5.1. On the lowest level of the hierarchy, there are the privates (Pvt Gump, Ryan and X). They are usually under the command of a sergeant (Sgt Drill and Horvath). Above sergeants, we find lieutenants (Lt Dan and Dewindt), commanded by the captain (Captain Miller). The typical order of command in the military is that soldiers at lower levels of the hierarchy report to one level above, while higher level soldiers give commands to one level below. For example, the path of some imaginary information from Pvt Gump to Pvt Ryan could be: (1) Pvt Gump reports to Sgt Drill, (2) Sgt Drill includes this information in his report to Lt Dan, (3) Lt Dan also includes the info in his report to Captain Miller, (4) the captain makes a decision and gives a command to Lt Dewindt, (5) Lt Dewindt commands Sgt Horvath accordingly, (6) Sgt Horvath then gives the corresponding command to Pvt Ryan. Such a path may describe a situation where Pvt Gump observes something important in the battlefield which should be reported to higher levels, from which the reacting commands seep down to the lower levels.

How does this regular path relate to shortest paths? In our imaginary organization, this regular path is also the shortest path, as we cannot find a path between Pvt Gump and Pvt Ryan with fewer steps. In fact, the organization is so simple that we only have one reasonable path between Pvt Gump and Pvt Ryan. All other paths will contain loops, meaning that there is at least one soldier that appears twice on the path. Let's make our organization a little more complex and realistic.

Consider that Pvt X is doing a special service for the military and spends half of his day under the command of Sgt Drill and the other half under Sgt Horvath (i.e., he is part of a liaison squad enabling communication between the units commanded

Fig. 5.1 Military hierarchy

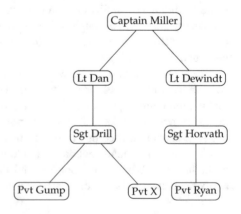

Fig. 5.2 Military hierarchy.
Shortest path vs. the regular
path

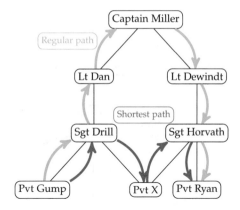

by the two sergeants). The network representation (see Fig. 5.2) of this modified
organization differs in only one edge going between Pvt X and Sgt Horvath. This
small modification, however, uncovers an interesting phenomenon. In the modified
network, the shortest path between Pvt Gump and Pvt Ryan is no longer through
lieutenants and captains. The regular path is unchanged, but the shortest path is Pvt
Gump → Sgt Drill → Pvt X → Sgt Horvath → Pvt Ryan. The corresponding story
could be that Pvt Gump reports to Sgt Drill, who orders Pvt X to report something
to Sgt Horvath, who gives the command to Pvt Ryan. This absolutely can be done
and fits within the norms of the army, but it is rather unusual. The shortest path
seems odd and breaches the everyday logic of the military network. We can say that
the shortest path is theoretically usable, but it seems practically non-traversable.
Moreover, in this case, the regular path coinciding with the internal logic of the
network is longer than the shortest possible path.

Besides the clear conflict between the shortest path and the regular path,
implementing the two paths will have different effects on the organization. By using
the regular path, high-level decision makers are notified about the event happening
at lower levels and can make use of this knowledge in later decisions. The usage of
the shortest path, while it enables a faster reaction, prevents the information from
escalating and leaves the army in a different state. Can such illogical but short paths
be used within the army under any circumstances? Are there unusual situations in
which the everyday practices can be overridden? Well, it depends on the level of
unusuality. Let's illustrate this with a short story about Hungary's participation in
the Second World War.

The participation of the Hungarian 2nd Army in the Second World War on the
side of Nazi Germany was undoubtedly surrounded by a great amount of unusuality.
There are many books[24] covering the stories of the battles facing Russian soldiers
on the eastern theatre of the war near the Don River. During its 12 months of activity
in 1942–1943 on the Russian front in the framework of the Operation Barbarossa,
the 2nd Hungarian Army's losses were enormous. Of an initial force of about
200,000 Hungarian soldiers 125,000 were killed in action, wounded or captured.
These losses were the result of the power of the Russian army, the extreme cold

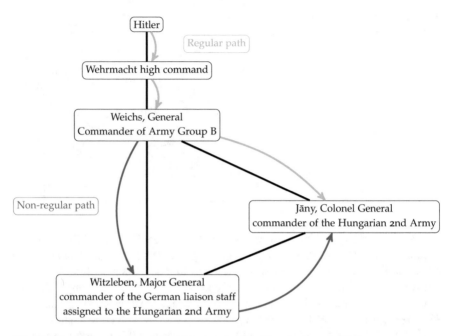

Fig. 5.3 Illustrative structure of the relevant parts of the German Army in 1943

and the commanding structure of the German Army Group B under which the Hungarian 2nd Army appeared as a sub-unit. The commander of the Hungarian 2nd Army was Gusztáv Jány (see Fig. 5.3). A German liaison staff commanded by General Hermann von Witzleben was assigned to work under the Hungarian army, coordinating the movements of the German and Hungarian armies. Thus his troops were somewhat subordinates of the two armies at the same time. The German Army Group B was commanded by Maximilian von Weichs, who received orders from the Wehrmacht high command and eventually from Adolph Hitler.

After a few months of Hungarians and Russians peering at each other on the banks of the Don, the Russians began their counter-attack on the front line of the Hungarian 2nd Army on 12 January 1943. During this attack, most of the Hungarian units were quickly encircled and either annihilated or forced to open terrain where they succumbed to the extreme cold. Facing the situation and the casualties Jány tried to obtain a command to withdraw his troops, using the standard chain of commands in the army. He sent messages to his superiors with the immediate request for retreat. After days of bloody massacre, the answer from the German high command remained the same: "In accordance with the Führer's decision, the positions ... must be held to the last man under all conditions". As a parallel action, on 15 January, Weichs asked Witzleben to meet Jány and to unofficially persuade him to order the immediate retreat. The reason for this unofficial action was that Weichs himself did not want to give a withdrawal order contradicting Hitler's instructions. On 17 January Jány eventually ordered his troops to commence retreating. It was only on 22 January when the headquarters of Army Group B decided, with Hitler's permission, to withdraw the Hungarian 2nd Army from the front line.

So what happened? Jány tried to use the regular path (chain of command) to report and react according to the decision of high-level decision makers. However, his situation was really unusual. He lost thousands of soldiers day-by-day and the regular path was too slow to properly react to the situation. While Weichs and other German commanders on site agreed with Jány about the immediate retreat, they didn't want to conflict with higher decisions. Instead, Weichs unofficially notified Jány through his German subordinate, by communicating that "high orders should always be interpreted in accordance with the situation". This act convinced Jány, that he really had his last men standing, so it was time to withdraw his troops. As a result, he retreated 5 days before the official permission. Those few days saved thousands of his people. In this case, the non-regular path was undoubtedly odd but worked and saved lives. How frequently do such events, that require non-regular paths, happen in the military? Well, horse sense suggests that if such events were prevalent, then the military would not work at all. So we should expect the great majority of paths to be regular and just a small portion of the paths to be non-regular.

The large scale Internet also possesses definite internal logic. It does in spite of the fact that it has a strong self-organizing characteristic in its evolution. Indeed, the Internet has grown into an intricate interconnection system through the uncoordinated process of nodes freely joining to the network. Although this joining process was truly without central coordination, it wasn't without laws. Relations between nodes have evolved to show a similar structure to what we saw in the military example. To shed some light on how this can happen, let's follow the imaginary story of the people of the little town Castle Rock, Maine.

During strong winters in Castle Rock, people are doomed to stay in their houses. One day lonely locals decide to connect to each other to communicate (e-mail, chat, video chat, etc.) by using their computers. Out of that purpose, they form a civil company that creates a local network across the whole village. Interestingly, almost in parallel, a similar series of actions takes place at the nearby settlement of Salem's Lot. No wonder that soon the two towns decide to establish a communication cable to connect to each other.

People seem to be happy for a while, but not much later it comes to the locals' knowledge that an English town, Dunwich, also built a local area network for similar purposes. Would it be possible to also connect Castle Rock to Dunwich? The distance seems to be too much for the poor little town. Fortunately an entrepreneur at the Main County Trans-Atlantic Co. (MCT) undertakes the task of building an underwater cable through the Atlantic Ocean and connects the MCT to Castle Rock and than to Dunwich. Thus providing long distance communication services to both towns for a monthly fee. Our newly born communication network, at this state, has four nodes: Maine County Trans-Atlantic Co, Dunwich, Castle Rock and Salem's Lot. The little town of Castle Rock initiated the building of a large-scale computer

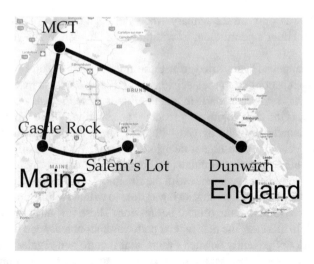

Fig. 5.4 An embryonic model of the Internet, where Castle Rock connects to nearby Salem's Lot directly and Dunwich in England through Main County Trans-Atlantic (MCT)

network, the nodes of which are networking companies. It is just like a tiny Internet (see Fig. 5.4).

Soon enough, however, conflicts arise when people at Salem's Lot take to regularly using the network spending most of their spare time communicating with the nice people from Dunwich! Notice that Salem did not spend money itself on building the network or buying the service from MCT, even so they can reach Dunwich using the networking resources of the nearby Castle Rock. But is it fair to Castle Rock to load its network, possibly slowing its communication service, due to the traffic of Salem's Lot? Through a connection that was initially established out of mutual agreement to exchange network traffic free of charge? Castle Rock pays Maine Trans-Atlantic to reach Dunwich, but it is free for Salem's Lot! The path from Salem to Dunwich does not seem to be regular at all! It is not in the logic of the network to use such a path. Castle Rock would surely cease the cooperation or ask for some compensation, thus making Salem's Lot a customer of Castle Rock.

Furthermore, a more interesting situation arises when the people of Castle Rock begin to feel unsafe about the single existing communication path to Dunwich. A Canadian company called Canadian Federal Communications (CF), noticing the business opportunity, comes forward also offering data transit services across the Atlantic for a fee. So, Castle Rock also chooses to connect to Canadian Federal as a backup plan for when the Maine Trans-Atlantic link goes down due to some error. Our tiny Internet has grown into a complex 5-node super-highway (see Fig. 5.5).

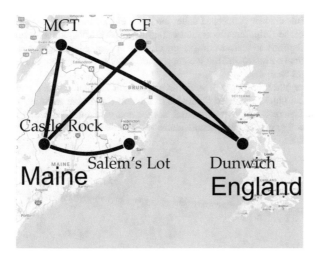

Fig. 5.5 A tiny model of the Internet initiated by the people at Castle Rock to communicate with the outside world by connecting nearby town Salem's Lot and Dunwich in England by using the transit services of Main County Trans-Atlantic (MCT) and Canadian Federal Co. (CF) as a backup route

All seems to be fine at the moment and will remain so as long as the Canadians do not decide to communicate with Maine County Trans-Atlantic, generating an even more serious conflict. Currently MCT and CF are only connected via their common customer's network. Should Castle Rock allow the CF to communicate with MCT? Absolutely not! It is not logical to provide transit services to them for free! The resources alone would not be enough to serve the transmission requirements of such large companies. Canadian Federal and Maine Trans-Atlantic should build their own network or use the transit services of even larger telecom service providers. It does not matter if the path connecting the two becomes longer as using the shortest path through a customer is not a viable solution! It is not considered to be regular!

As we can see the labyrinthine network, which we now call the Internet, connecting the telecom companies of the world also has some structure to it and works along some rules. Similarly to the military system, it contains a communication hierarchy formed by customer-provider relationships generating strict communication policies governed by complex business interests.

Is it possible that other networks also have similar internal logic? Is it possible that similar reasons can make paths longer than the shortest path? It is time to not just philosophies about real paths, but to do some ground truth measurements and see what phenomena are actually supported by real data.

Chapter 6
On the Trail of Nature: Collecting Scientific Evidence

Over every mountain there is a path, although it may not be seen from the valley.

— Theodore Roethke

To get closer to understand the nature of paths we need two kinds of data about the same networked system like the Internet, or the Bridges of Königsberg. First, we need at least an approximate network connecting its nodes and a large number of paths collected from real traces of packets. Using the words of the Bridges of Königsberg problem, we need the network representation of the lands and bridges (Fig. 3.2) and the footprints of people's afternoon walks.

In the last two decades, the flurry of network science [2] in all fields (biology, physics, sociology, technology) has resulted in the reconstruction of thousands of networks lying behind real-world systems [9]. Ranging from the classic Kevin Bacon game[1] over the network of Holywood movie actors, through the metabolic and social networks to the sexual contact of people, we now have systematically collected well-organized and publicly available data repositories about real-world networks (e.g., SNAP [17]). So downloading and computing something interesting over the network representation of cell metabolism in our cells is now an afternoon of laboratory work for an undergraduate student. What about the paths? Well, gathering paths seems to be a very different task compared to inferring simple connections in a network. The techniques working for the identification of network edges are generally not usable for gathering paths. Recall the Bridges of Königsberg again as an illustration. The map of downtown Königsberg is easy to get. Just jump into a map store and buy one, or draw an approximate map after a few days of walking in the streets of the city. The map or the network of the city is a form of public information. But what about the paths? Well, the paths belong to people. The paths describe the habits of people and tell us about them. About their favorite

---

[1] https://oracleofbacon.org/.

A. Gulyás et al., *Paths*, https://doi.org/10.1007/978-3-030-47545-1_6

places, the location of their homes and even about their health (if they prefer long or short walks). The nature of paths seems to be somewhat confidential. Some people may talk about it and give their names, others may talk about it anonymously and others may ignore you if you ask them about their paths.

Although gathering information about paths is not particularly easy, it is not hopeless either. Now we present four very different systems for which both the network data and the path data can be obtained to an appropriate extent. Our collection here will be based on the recent study of Attila Csoma and his colleagues about paths [6].

6.1   Flight Paths

When you board a plane of some airlines and get seated, you can find many things stuffed into the rear pocket of the seat in front of you. There are life-saving instructions, maps of the aircraft with the locations of exits, a sanitary bag, but there are also airline's magazines. In these magazines, among the advertisements about the most attractive flight destinations, there are usually nice maps showing all the flights operated by your airlines. If we could collect all magazines from the back pockets of all airlines, then we could easily reconstruct the flight network of the world, by considering the airports as the nodes of our network and the flights between them as the edges, no matter which airlines operate them. Although it would be quite time-consuming, it is absolutely doable.

Fortunately, there is a much simpler method for constructing the flight map of the world. Since flight information is public, there are public online data repositories which accumulate all the information about the flights all over the world. For example, the OpenFlights [21] project collects such data and makes the whole database publicly accessible. By listing, for example, all the flights of US airlines, the flight map of the US can be drawn (see Fig. 6.1).

Therefore, the reconstruction of the flight network is not rocket science, having the online datasets at hand. What about the paths? Well, a path is the multi-flight travel of somebody between the departing and destination airports through the flight network. Having a path means that we know the detailed flight information including the flight transfers for a given passenger or a set of passengers. Path information reveals how people choose transferring options at various airports, in cases where there is a lack of direct flights between the source and destination. Knowing a large number of paths is equal to knowing all the transfers of passengers for their trips, which is not something we should know without their consent and as such, there are no online databases for them. How can we obtain paths then? Well, we can take an "indirect" path to these paths. There are various flight-trip-planner portals that offer tickets between arbitrary sources and destinations all over the world. On such websites we can plan our whole journey and buy the tickets online. We can safely assume that many of the passengers buy their tickets using similar websites. So, what we can do is observe flights between randomly picked

Fig. 6.1 The flight network of the US

airports and consider these offerings as paths that passengers could really choose for their trips. Gathering thousands of such flight offerings can give us a fairly usable approximation of paths used by real people, without raising confidentiality issues. Note that a particular trip can be the result of intricate business interactions among many different airlines and the passenger itself, so similarly to the Internet, the airport network also seems to be working without central coordination. What expectations may we have about the paths coming out of such a network? For starters, consider the position of Hungary's flagship airlines (which has stopped flying recently) in the flight system of the whole world.

MALÉV Hungarian Airlines was the principal airline of Hungary from 1946 to 2012. It had its head office in Budapest, with its main operations at Budapest Liszt Ferenc International Airport. In its best years, Malév operated direct flights between Budapest and New York, and Budapest and Moscow. In this respect, Moscow and New York could be connected by a path of length two through Budapest. What can we say about this two-step path offered by Malév? Well since one could travel from Moscow to Budapest and from Budapest to New York with Malév, we must say that this path is usable by passengers. However, Moscow and New York are huge metropolises with airports serving around 30 and 60 million people respectively, while Budapest airport is used by around 10 million people in a year. So, connecting these cities through the relatively small airport of Budapest looks a bit odd and we may suspect that the great majority of people would use other airports (e.g., Heathrow, Charles de Gaulle, Schiphol or Frankfurt) for changing flights between Moscow and New York. Although the background and the way of operation of the

flight network are very different, we can suspect a similar underlying hierarchy of
the airports as we have seen in the case of the military or the Internet.

6.2   Paths from a Word Maze

Word games are fun and entertain people regardless of their age. The Last and
First game, for example, is frequently played between children and parents or
grandparents. The essence of the game is to say a word which begins with the final
letter of the previous word. For example, the word chain camel → lion → napkin
→ nest → tiger → raven can be the result of an afternoon game between granny
and grandchild. Wait a minute! Doesn't this look like a path? A path of words? Sure
it is! But instead of leading to somewhere, the aim of this chain is to go on as long
as granny is awake and the grandchild is not bored. In this respect, the game does
not have a destination (at least in terms of words). But can we twist the game a little
bit so that the word paths will lead somewhere? Word ladder games are designed
just for this purpose.

In a word ladder game, players navigate between fixed length source and destination words step-by-step by changing only a single letter at a time. For example, the
word path fit-fat-cat is a good solution of a game with source word "fit" and target
word "cat". This path is now very similar to our flight paths, in the sense that they
have a definite source and destination and "transfers" can be made between words.
Is there a public repository accumulating solutions of word ladder games played
by people? Well, luckily there is[15]. Recently, Attila Csoma and his colleagues
have developed a word ladder game for smartphones in a framework of a scientific
project, and collect the word paths of people. After the users install the game, they
are asked to transform a randomly picked three-letter English source word into an
also random three-letter target word through meaningful intermediate three-letter
English words by changing only a single letter at a time. The word paths entered
by the users are collected anonymously. Fortunately, word path game solutions do
not seem to be as confidential as flight information, as hundreds of users shared
thousands of word paths (despite the clear deficiencies of a game developed by
university researchers). These paths can be considered as the footprints of humans
navigation over the word morph network of the English language.

More specifically, the collected paths are footprints of the process by which
people master their navigational skills in the network lying behind the game. The
word morph network is a network of three-letter English words, in which two words
are connected by an edge if they differ in only a single letter at the same position
(see Fig. 6.2). For example, the word "FIT" is connected to the word "FAT" as they
differ only in their middle letter. "FAT" is linked to "CAT" as they differ in their
first letter, but "FIT" and "CAT" are not connected in this network since they differ
in more than one letter. The paths collected from players are paths in this network
and reflect valuable information about how people try to navigate between nodes.

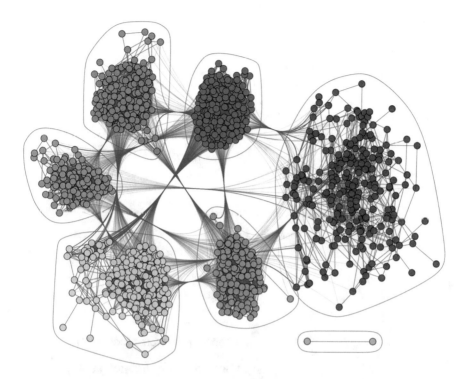

Fig. 6.2 The word morph network is a network of three-letter English words, in which two words are connected by a link if they differ only in a single letter. For example, "FIT" is linked to "FAT" as they differ only in the middle letter, but "FIT" and "CAT" are not neighbors in this network since more than one letter differs in them

Figure 6.3 shows a small portion of the word morph network and illustrates two solutions for a puzzle between source and target words "YOB" and "WAY".

What can we expect from these word paths? How will they look? Will we find "odd" paths and "regular" paths similarly to the chain of commands in the military and flight paths? As a sanity check, we present a common finding the players reached after playing some games. They realized that words are not equal in this game and some words can be used for various functions. The most basic puzzles, like the "FIT" → "CAT" one, can be solved by simply getting closer and closer to the destination in terms of matching letters. In "FIT" there is one matching letter with "CAT", in "FAT" there are two matching letters, while in the destination "CAT" all letters are matched. How about the "TIP" → "ALE" puzzle? This is much more complicated since the consonants and vowels are at completely opposite positions. In this case, the above strategy simply doesn't work. Now the players have to find words with back-to-back consonants or vowels, where such letters can be swapped. For example, the "TIP" → "TIT" → "AIT" → "ALT" → "ALE" is a solution, where the intermediate word "TIT" is just there to turn to "AIT" at which vowels are back-to-back at the front, which then can be changed to "ALT" at which consonants are

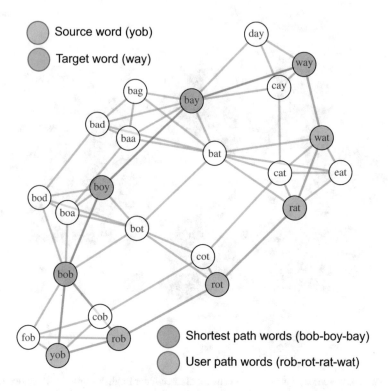

Fig. 6.3 A word morph game example with source and target words "YOB" and "WAY". A shortest solution is displayed in red, while a solution given by a specific player is shown in green

back-to-back at the end and is just one step from "ALE". People quickly memorize such "trade" words like "AIT" or "ALT" and reuse them in further puzzles. So, it seems that there is also some underlying logic in this simple word game. But is this logic similar to what we can find on the Internet or in the flight network?

6.3 Internet Paths

The reconstruction of the network to which the Internet has evolved after more than three decades, grasped the attention of many researchers worldwide. As the Internet is built over electronic devices, its topology could be reconstructed by collecting all the connection-related data residing in each of its constituting nodes (i.e., computers). However, unlike the airport network where the flights between airports is public information, the connection information between Internet providers is not easy to obtain. The traffic agreements between internet providers are usually kept confidential. Therefore, it seems that we cannot even get the underlying network of the Internet in a straightforward way, not to mention the paths we are curious about. It turns out, however, that some Internet hacks can help us find the paths. In this respect, the Internet is a unique platform for researching paths.

To get a picture of how packets go through the Internet, we have to understand some fundamentals of computer networking first. A packet traveling between computers is nothing more than a few bits of information encoded as electronic signals. Every packet has a source and destination address and a payload which should be delivered to the destination. Packets usually do not change and do not think, which is in high contrast with the people walking through the bridges of Königsberg. Instead of the packets, the computers (i.e., the lands) "think". What does a computer (let's refer to them as nodes in a network context) do when receiving a packet? First, it looks into the destination address. If the destination address is the current node, then it "consumes" the packet. After extracting the payload data, the packet is destroyed. If the destination address is not the node receiving the packet, it has to find out how to forward the packet to its destination. Who tells the node how to find this out? People! Not ordinary people of course, but networking people whose job it is to operate networks. There is a routing table in every node, which is very similar to road signs. It indicates the next turn a packet should take on the way to a specific region of the network.

Consider Fig. 6.4 as an example. There are seven computers marked with letters (A, B, C, D, E, F, G) forming a very simple network of seven edges. Now suppose that D wants to send a packet to G. As a first step, it creates a packet containing the source address D, the destination address G and also the payload data, just like a postal letter. Now node D has to send the packet to G. The situation of D is extremely simple as it has no choice where to send the packet, its only option is B. Quite the contrary, at node B (as it is not the destination) there are plenty of options to forward to. How will B decide? Well, a capable network operator configured B to solve such situations. The operator creates the routing table of B, from which B can read the next step of the packet destined to G: it must be given to C. Similarly, C is instructed to send packets with destination address G, to node G. As a result finally G receives the packet successfully. The full operation can be made according to the routing tables in the nodes (see Table 6.1). From these routing tables, all the paths

Fig. 6.4 A simple network of computers

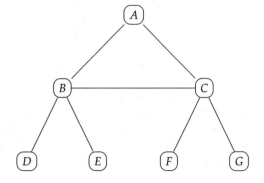

Table 6.1 Possible setting of routing tables for the network in Fig. 6.4

Routing tables of the nodes		
Node	Destination	Next step
B	D	D
	E	E
	A	A
	C, F, G	C
C	F	F
	G	G
	A	A
	B, D, E	B
A	B, D, E	B
	C, F, G	C
D	ALL	B
E	ALL	B
F	ALL	C
G	ALL	C

between any pairs of nodes can be reconstructed. The problem is that the nodes of the Internet belong to various networking corporations, which do not intend to disclose the routing tables. So, collecting the paths in such a way is not an option. What can we do then to get our paths?

Fortunately, there is something in computer networks which all networking companies are scared of. So scared, that they implement several mechanisms to detect and avoid them. These daemons are called loops. Consider that in our simple computer network in Fig. 6.4, every node is administrated by a distinct company, i.e., a different operating person. Consider that the operator of A notices that its direct connection to C is weak, e.g., it provides a slow connection. So the operator of A sets the routing table in A to forward every packet destined for C, F and G to B, avoiding the laggy direct connection to C. Independently, B also considers its connection to C as pretty weak and forwards all packets heading to C, F and G to A (see all the modified routing tables in Table 6.2, with the modifications shown in boldface).

Now, what happens with the packet destined for G after these tiny, uncoordinated modifications in the routing tables? Well, it starts at D as seen before. B sends it to A according to its routing table, but A sends it back to B, B sends it back to A, A sends it back to B ... and so on forever. There is an infinite loop between B and A. After some time, node B and A are only occupied by looping the packet infinitely, which eats their resources, pointlessly generates a lot of heat, and most importantly ruins the operation of the whole network. You may think that the routing settings are carefully negotiated between networking operators, so such things could not happen. Well, routing settings are usually well negotiated, but the human factor is always there. Misconfigurations happen every day on the Internet. One of the most famous examples was in 2008 when due to a routing table misconfiguration in a

Table 6.2 Setting of routing tables leading to a loop for the network in Fig. 6.4

Routing tables of the nodes		
Node	Destination	Next step
B	D	D
	E	E
	A	A
	C, F, G	A
C	F	F
	G	G
	A	A
	B, D, E	B
A	B, D, E	B
	C, F, G	**B**
D	ALL	B
E	ALL	B
F	ALL	C
G	ALL	C

node of the Pakistan Telecom, a large portion of YouTube's traffic was hijacked and discarded in Pakistan.

Thus, loops are dangerous things in computer networking and one should immediately detect and avoid them. The current solution for that is to include so-called time-to-live (TTL) information in the packets. This is a simple number which is decremented by every node the packet visits. If this number becomes zero, the packet will be destroyed even if it hasn't reached its destination. This way, bad configurations will have limited effects as packets cannot travel for an infinite time between the nodes. When a packet's TTL becomes zero and it is not at its destination, that is a good sign that something is wrong with the network. In this case, the node which destroys the packet sends an alert to the source address found in the packet stating that something is wrong. And this is where the approximate tracing of packets becomes possible.

Consider the following hack. We want to know the nodes on the path towards a destination node D from source node S. First, we send out a packet from S and set its TTL value to 1. This packet will reach one of S's neighbors (let's say A), which will destroy the packet and notify S that something went wrong. From this notification, we record, that our packet has visited node A. Now we start again and send out the packet, but this time setting its TTL to 2. The packet would not be destroyed by node A as its TTL becomes 1 when A decrements it. So A will forward the packet to somewhere, let's say to B. Since at B the TTL is decremented again, it becomes zero, so B destroys the packet and sends back a notification to S that something went wrong. At S, we record that the packet has visited B so the path to B is $S \rightarrow A \rightarrow B$. The process continues until a large enough TTL setting lets our packet reach its destination. Sounds a bit complicated but this is all we have. This method (called traceroute) gives us approximate paths and we can use this method from any node connected to the Internet, even from your laptop. Fortunately, there are public datasets which contain such Internet paths collected from thousands of different locations. These datasets (see for example the website of the Center for Applied Internet Data Analysis[4]) can give us millions of paths from which an approximate map of the Internet can be recovered.

How can we construct the topology of the network from paths? Suppose that we have three paths: *Path 1*: A → B → C, *Path 2*: A → B → D → E, *Path 3*: E → C → F → A. By analyzing *Path 1*, we see that there are nodes A, B and C and there are edges between A and B, and between B and C. From this, we can draw a network shown in Fig. 6.5.

Now we analyze *Path 2*. We realize, that there are also nodes D and E in the network, and we locate two edges: B → D and D → E (see Fig. 6.6). Finally, the observation of *Path 3* adds a new node F and three edges: E → C, C → F, and F → A. So, after processing the three paths, we get the network in Fig. 6.7. After processing more and more paths, we will have more and more appropriate pictures of the whole network.

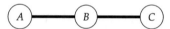

Fig. 6.5 Constructing a network based on its paths, Phase 1

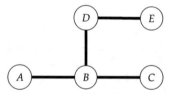

Fig. 6.6 Phase 2

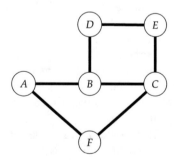

Fig. 6.7 Phase 3

6.4 Paths from the Human Brain

The human brain is one of the most complex networks one could imagine. Understanding even parts of its functionality is extremely challenging and is still one of the biggest mysteries of human life. Here we are interested in the paths inside the brain: the paths over which information can travel between different parts of the brain. Getting realistic paths from inside the human brain is extremely hard, if not impossible. As a consequence, almost all current studies concerning path-related analysis simply assume that signaling uses shortest paths, meaning that we suppose brain signals follow the shortest possible path in the brain. Similarly to these studies, we have to accept that we cannot get paths out from the brain in a direct manner. In the case of the Internet and the flight network, the confidentiality of the path related data was the main obstacle of getting direct paths. In the case of the brain, we just simply don't have the appropriate technology (yet) which could identify the paths for us. What can we do then? Is there a similar hack for the brain that we used over the Internet? What kind of data is currently available about the flow of information inside the brain? We will go through these questions in the following paragraphs.

The Human Genome Project (Fig. 6.8) was one of the biggest endeavors of mankind and was surrounded by the most remarkable scientific collaboration across many nations. Its target was to determine the sequence of nucleotide base pairs that make up the human DNA. Upon its completion, at a press conference at the White House on the 26th June 2000, Bill Clinton evaluated the resulting map of the human genome as: "Without a doubt, this is the most important, most wondrous map ever produced by humankind."

A similar endeavor started in 2011 when the Human Connectome Project was awarded by the National Institutes of Health. This project is targeted to construct the "map of the brain", i.e., to discover the structural and functional neural connections within the human brain. The structural map means that we locate specific brain areas (these will give us the nodes) and the physical connections (which will give us the edges) between them. How can one do this without slicing up somebody's brain? Well, this is what the "non-invasive" brain mapping methods are used for. With a quite complicated method called DSI (Diffusion Spectrum Imaging), the diffusion of water molecules can be observed inside the brain. To get a picture of how DSI works, think about constructing a road network by observing only the movement of cars at various observation points throughout the area you want to map. You cannot see the roads themselves, but you can see the cars at these observation points and you can write the direction and intensity of their movements. By collecting all this information from the observation points, after some non-trivial computerized post-processing, we can create an approximate map of roads and cities in the given area. Interestingly, the process is very similar to the operation of WAZE, a popular navigation software (now owned by Google), where the positions of WAZE users are collected in an anonymized database. In this case, however, the exact map is drawn by volunteer editors, using the draft map deduced from the database. In DSI, the cars

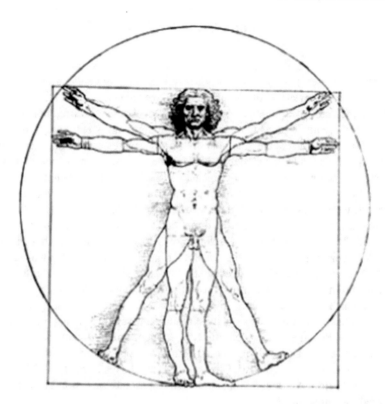

Fig. 6.8 The Vitruvian Man depicting normal human body proportions is often used to symbolize The Human Genom Project as Leonardo da Vinci created it in 1490, exactly a half a millennium before the project began in 1990. [Public Domain; Leonardo da Vinci via Wikimedia Commons]

are water molecules, which are observed at various points in the brain by using MRI (Magnetic Resonance Imaging) devices. A picture about the human connectome, i.e., an approximate picture of one's map of neural connections in the brain, obtained via DSI, can be seen in Fig. 6.9.

Thanks to DSI we can have one's connectome, i.e., we have the network over which our paths form. What can we say about the paths? It seems that at this time we can say nothing about them in a direct manner. But there is something we can do to at least estimate brain paths better than simple shortest paths? fMRI[20] (functional Magnetic Resonance Imaging) is a method with which one can reason about brain activity. With fMRI, the blood oxygenation of various regions in the brain can be measured. Since blood flow and oxygenation are correlated with brain activity (active brain regions use more energy and require a higher level of oxygen in the blood), the changes in blood oxygenation reveal the neural activity. Back to our city-roads-cars analogy, fMRI is quite similar to the task of reasoning about the operation of a city, by observing the density of cars in its various districts.

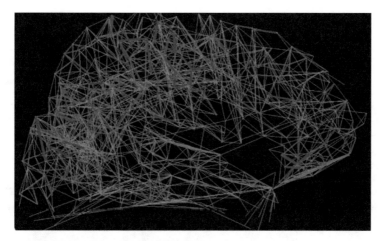

Fig. 6.9 The human neural network in the brain reconstructed via DSI, from Patric Hagmann et al. "Mapping the structural core of human cerebral cortex". In: PLoS biology 6.7 (2008), e159

How can we approximate paths in the brain? Well, DSI delivers an approximate "network" of the brain, meaning that it gives us the nodes and the physical connections (the bridges in the Königsberg analogy) between them. The fMRI gives a different "network" in which brain regions are not physically, but functionally or logically connected, meaning that they frequently act together, so they seem to implement similar functionality. Can we make use of some trick and infer something path-like from these data? Here is what we can do. By combining structural (DSI) and functional (fMRI) data, we estimate paths through which neural signals might propagate using the following hack. First, we have to identify the sources (i.e., the starting node) and destinations (the nodes where the path ends) of our paths. From the fMRI signals, we can identify brain regions, which frequently exhibit neural activity at the same time. Simultaneous activity hints that these brain regions are working on the same task and are likely to exchange information in the form of neural signals. We identify these simultaneously active brain regions as the source-destination pairs of our paths. Now we have to figure out the path between these sources and destinations. In cases where there is a lack of information, we could determine the shortest path between the endpoints of our paths using, for example, Dijkstra's algorithm over the structural connectivity network obtained from DSI. Figure 6.10 shows an illustrative brain network of 15 nodes. Over this network, we would like to approximate the possible signaling path between regions 1 and 15. The shortest path approximation will give the $1 \to 5 \to 12 \to 15$ path for this. In fact, most studies in the related literature use this simple approximation.

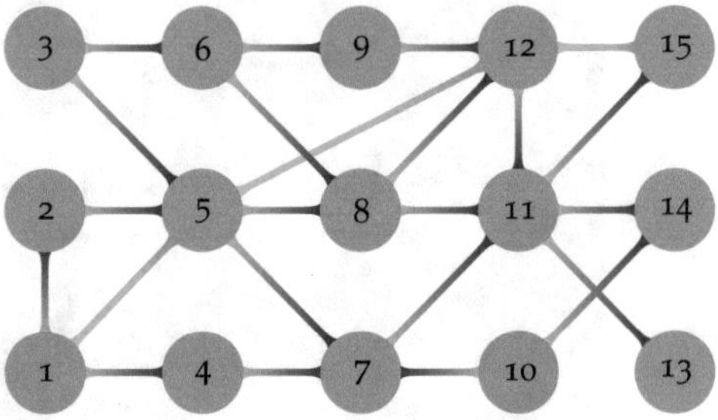

Fig. 6.10 Inferring path from
the human brain using the
shortest path assumption

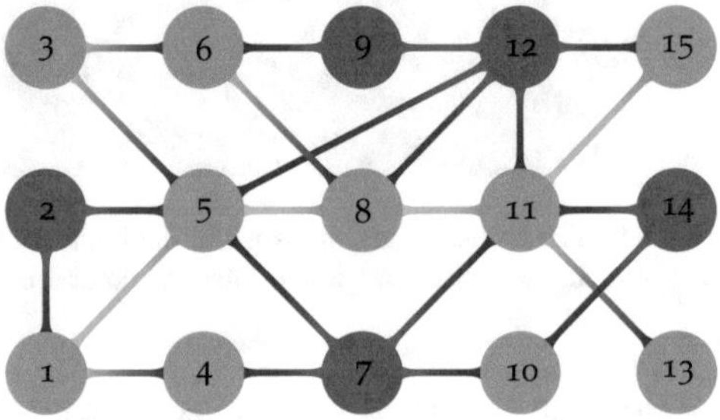

Fig. 6.11 Shortest path over
the active subnetwork at a
given time instant

Due to the extreme complexity of the brain, as of now, we do not have direct
information about the paths inside, but we can do slightly better than simple shortest
paths. We can use the fMRI to identify regions with neural activity and from the DSI
network, we can exclude the inactive regions during signal transmission between the
endpoints of our paths. We can do this because inactive regions are not likely to pass
on any information. By excluding inactive regions, we will get the active subnetwork
for every information exchange we are curious about. Figure 6.11 shows the same
network we can see in Fig. 6.10, but the red regions (2,7,9,12,14) are inactive, and
thus are excluded from the path approximation. Therefore, we will find the shortest
path between 1 and 15, but we cannot step onto the red regions. The shortest path in
this new scenario is $1 \rightarrow 5 \rightarrow 8 \rightarrow 11 \rightarrow 15$, which is longer than the shortest path
in the original DSI network.

While we cannot validate with empirical data whether these paths (see Fig. 6.12)
are actually used for the flow of neural signals, we can at least consider these paths
as lower bounds on the length of the real brain paths.

Fig. 6.12 Empirical paths in
the human brain

Chapter 7
The Universal Nature of Paths

The way is like the bending of a bow. To achieve its ends the top must bend down and the bottom rise up.

— Tao Te Ching LXXVII

Although it was not always particularly straightforward, by the end of the last chapter we came up with methods enabling the exact measurement or the estimation of *empirical* paths in various real-life networks. From now on, we will refer to the paths coming from measurements in real networks as empirical paths, to clearly distinguish them from other paths (e.g., shortest paths) with which we will compare them later. Before taking a look at the properties of the empirical paths, let's take some time to overview some numbers about our networks and paths (see Table 7.1).

The first row of Table 7.1 shows the number of nodes in each network. We can see that those networks are not just from completely different corners of life, but their sizes also vary significantly. In the case of the Internet, it contains more than 50 thousand nodes, while there are only 1015 three-letter English words constituting the word morph network. The second row presents the number of edges in each network. Read with the node sizes, we can conclude that these networks are much bigger than the network behind the seven bridges of Königsberg problem (which had only four nodes and seven edges, see Fig. 3.2).

The third-row reports on the so-called diameter of the networks, which is the longest among the distances of any two nodes. Remember that the distance is the shortest path among two nodes. To understand the concept, we can take, for example, the diameter of the Universe as the distance of the two galaxies that are the farthest away from each other. In that case, the distance is measured with the shortest possible straight line through free space. Measuring the distance in a network of course is done by counting the number of links from node to node. The diameter of the Königsberg network in Fig. 3.2 is two, as no two nodes are farther away from each other by walking the shortest path. Interestingly our networks, although larger than the Königsberg network by orders of magnitude, have an extremely low diameter. This property, that the diameter can be very small despite the network being very large is also known as the small world property [25], which most of the real networks readily exhibit. To intuitively grasp the small world property, think

© The Author(s) 2021
A. Gulyás et al., *Paths*, https://doi.org/10.1007/978-3-030-47545-1_7

Table 7.1 Basic properties of our networks and paths

Network	Airport	Internet	Brain	Word morph
Number of nodes	3433	52,194	1015	1015
Number of edges	20,347	117,251	12,596	8320
Diameter	13	11	6	9
Average shortest path	3.98	3.93	2.997	3.52
Number of emp. paths	13,722	2,422,001	394,072	2700
Average empirical path	4.67	4.21	4.16	3.82

about the friendship network (e.g., Facebook) of people around the world. Although there are billions of people in this network, any two persons can be connected by using a friendship path of around six people. A friendship path starts with some guy and goes on to one of his friends, then to one of his friend's friends, then to one of his friend's friend's friends, and so on. The small world phenomenon is frequently illustrated by the popular term "Six degrees of separation" [13, 14] used in John Guare's play (Fig. 7.1), in which Ouisa Kitteridge says: "I read somewhere that everybody on this planet is separated by only six other people. Six degrees of separation between us and everyone else on this planet. The President of the United States, a gondolier in Venice, just fill in the names. I find it extremely comforting that we're so close."

The final metric belonging to the networks presented in the fourth row of Table 7.1, is the average distance between their nodes. This means, that we compute the lengths of shortest paths (e.g., by using Dijkstra's algorithm) between all possible pairs of nodes and then we take the average of all these lengths. This will give a smaller number than the diameter (which is the maximum among the shortest paths) and is remarkably similar for all our networks. For the Königsberg network in Fig. 3.2, the average distance is 1.16667.

Regarding the empirical paths, we have two rows in Table 7.1. The fifth row presents the number of paths we have been able to collect by our measurement hacks in various networks (ranges from several thousand to millions, in the case of the Internet). The final (sixth) row shows the average length of our empirical paths given by the traceroutes over the Internet, ticket bookings over the airport networks, path estimations in the brain, and puzzle solutions in the word-morph game. We can see that the average empirical path is longer than the average of the shortest paths (distances), which insinuates that nature does not always use the shortest possible path over its networks, not even in networks where the shortest path could easily be found. Although the difference is not extremely large, it is not negligible, especially compared to the length of the paths (3–4). So it seems that real empirical paths are 10–30% longer on average than the shortest paths.

Now recall our introductory examples! The tale of the little cock, the users of the open proxy system and the mind map presentations. Our impression about these examples was that paths used in real life may be somewhat longer for some reason than the shortest possible path. This impression about the presence of detours is now

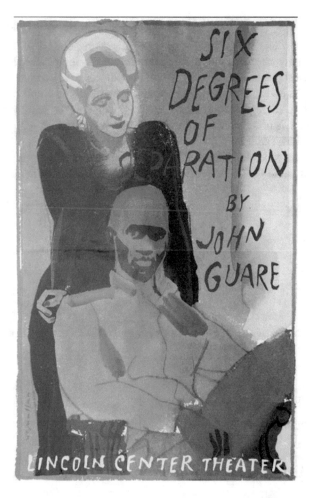

Fig. 7.1 Six degrees of separation. The poster of the play created by James McMullan. [With the permission of James McMullan]

confirmed by real measurements in four networks having very diverse backgrounds. In short, not just people, but many other things seem to favour detours. But is it just the average length of the paths that exhibit similarities or is there more in common? Let's continue with a bit deeper examination of the length of the empirical paths and the possible path selection rules used by nature.

7.1   Rule 1: Pick a Short (But Not Necessarily Shortest) Path

Let's define a metric which can show to what extent the empirical paths are longer than the shortest possible path. We will call the difference between the length of the empirical paths and the shortest path as the "stretch" of the empirical path. In the example of Fig. 7.2, the shortest path between A and C is the green path (of length 2). The red path is of length three, thus it will have a stretch of $3 - 2 = 1$. Similarly, the blue path has a stretch of 2.

Now let's see what percentage of the empirical paths exhibit a stretch of zero (i.e., the empirical path is the shortest path), stretch of 1, 2, 3, etc. Figure 7.3 depicts a

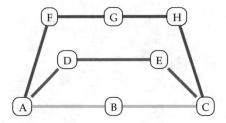

Fig. 7.2 The illustration of path stretch. The green path is the shortest, while the red and blue paths has a stretch of 1 and 2 respectively

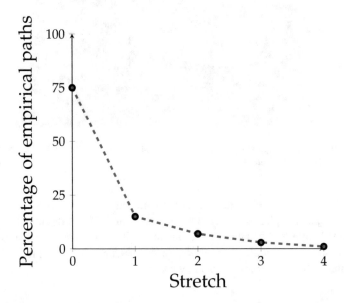

Fig. 7.3 A simplified sketch on the measured stretch of the paths relative to the shortest one found in our real-life systems. While most of the empirical paths exhibit zero stretch (confirming the shortest path assumption), a large fraction (20–40%) of the paths is "inflated" even up to 3–4 steps. The plot appropriately represents the distribution of path stretch that is found to be stunningly similarity in all four previously presented networks

simplified sketch of the summarized findings in our four real life networks showing the percentage of empirical paths as the function of stretch. Remarkably, all of our networks show very similar behavior in that regard. As the stretch increases, the percentage of empirical paths having that particular stretch decays pretty similarly. This means that it is not just the average stretch which is similar in real networks and paths, but in each network, we seem to find paths of a particular stretch with a similarly decaying chance. The overall behavior is also interesting. While around 60–80% of the empirical paths have zero stretch, the remaining paths exhibit stretch which can exceed up to 3–4 steps, or even more in some networks. From this result, two things follow. First, the plot confirms the efficiency of nature in the sense that most of its paths are shortest indeed. In this respect, nature definitely "prefers short paths". However, a non-negligible portion (20–40%) of stretched paths suggests that there may be other considerations when paths are picked in real networks. What kind of path selection rules produce similar results regarding the stretch of the paths? What are the guidelines when picking a path? For understanding this, we have to recall our main ideas about the internal logic of networks.

7.2 Rule 2: Use Regular Paths

We have seen before that there can be some kind of internal logic in networks, in the form of various hierarchies, which can affect the structure of paths. In case of the army example, this is quite obvious as the army is a fully hierarchical organization. In the case of the Internet or the air transportation system, a similar hierarchy can be reasoned; however, their presence is not so obvious. In the case of the human brain or the word-morph game even, reasoning about hierarchies behind the network seems non-trivial at this time.

How can we check if the stretch of the paths has something to do with these underlying hierarchies? How can we prove that the reason of an empirical path being slightly longer than the shortest path is to match the internal logic of the network? How can we define the hierarchy that can be used for all of our networks in the first place? A possible resolution to this problem is to use the so-called closeness centrality number of the nodes as the measure of hierarchy level. The closeness centrality, or centrality in short, of a node can be obtained by taking the number of nodes in the network except the node itself and dividing it by the sum of the lengths of the shortest paths from the node to every other node. Notice that the number is higher for nodes located more centrally in the network.

For our military example (with an extra lieutenant added) in Fig. 7.4, for Captain Miller we have $1 + 1 + 1 + 2 + 2 + 3 + 3 + 3 = 16$ as the sum of the lengths of the shortest paths to the others, and 8 as the number of nodes in the network except Captain Miller. Thus his centrality is $8/16 = 0.5$. Computing the centrality of the other soldiers as well (see Fig. 7.4), we get a clear reflection of the military hierarchy. The nodes in the inner part of the network with higher centrality can be

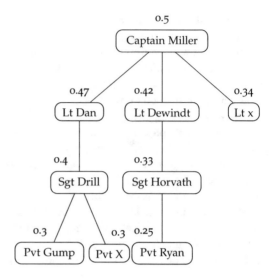

Fig. 7.4 Military hierarchy
with 3 lieutenants

considered as the core and the nodes with lower centrality as the periphery of the network.

By assigning a number to every node of a network reflecting the position in the hierarchy, we get their role in the internal logic of the network. Now, the question arises as if our empirical paths in the different networks have anything to do with those numbers. After analyzing our paths, we find that most (around 90%) of the empirical paths do not contain a large-small-large pattern forming a "valley" anywhere in their centrality sequence. For example, the path from Sgt. Drill towards Lt Horvath trough Lt Dan, Captain Miller and Lt Dewindt in Fig. 7.4 has a centrality sequence of $0.4, 0.47, 0, 5, 0.42, 0.33$, which contains no large-small-large patter in it (no "valley"). However, would there be a link between Pvt X and Sgt Horvath, the path from Sgt Drill towards Sgt Horvath through Private X would have a centrality sequence of $0.4, 0.3, 0.33$, containing a "valley".[1] The fact that the probability of finding such valleys in the empirical paths is very low suggests that in real networks higher level nodes do not prefer the exchange of information through their subordinates even if there are short paths through them. On most of the empirical paths, the centrality increases monotonically at first (upstream), or in other words goes "deeper into the center of the network", then starts to decrease (downstream), going "out of the network", until it reaches the destination. Or in other cases, the path goes upstream or downstream all the way. So the empirical paths coming from our measurements seem to follow the underlying hierarchy of the network. In other words, almost all empirical paths follow the internal logic of

---

[1]The watchful reader may argue that adding a new link to Pvt X would change also his centrality in the network (in our case, increasing his centrality above even that of his direct superiors), however, this odd artifact would diminish fast as new soldiers were enlisted in the army. For the sake of keeping our network example perspicuous, we omitted this correction here.

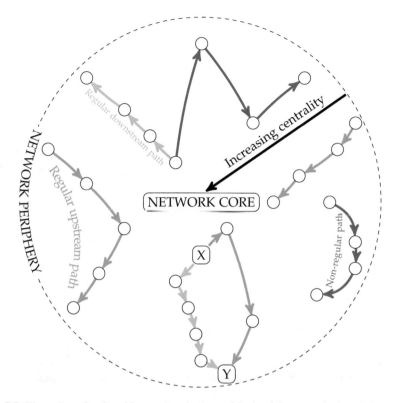

Fig. 7.5 Illustration of paths with regard to the internal logic of the network. A path is regular if it does not contain a large-small-large pattern forming a "valley" anywhere in its centrality sequence (green and orange paths). Red paths show examples of non-regular paths. An upstream path contains at least one step upwards in the hierarchy of the network (orange paths), while in downstream paths, the centrality decreases all the way (green paths)

the networks; they are "regular" by following the chain of commands. Figure 7.5 graphically illustrates such paths, where regular paths are colored green or orange. Now we can recall our example of the Hungarian 2nd army. Then we settled on the horse-sense conclusion that the great majority of paths were regular and we expected only a small subset of the paths to be non-regular. Well, measurements in this section now quantify the "great majority" as 90% and confirm our expectations.

7.3 Rule 3: Prefer Downstream

All right, so empirical paths follow the internal logic of the network even though it produces slightly longer paths. Can we say anything else? Well, there can be subtle differences between regular paths of similar length. For example, a path can contain upstream then downstream steps or only downstream steps. Recall that an upstream

Fig. 7.6 Military hierarchy:
downstream and upstream
paths

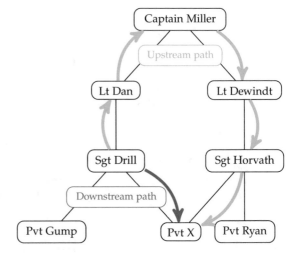

step goes upwards in the hierarchy, while a downstream step goes towards the
periphery of the network (see Fig. 7.5). Is there a preference among those? Should
military sergeants turn towards their commander, or can they give orders directly to
the units under their command? If we ask the question in such a form, the answer
seems pretty clear: sergeants surely can issue orders directly to their units. Thus, a
military sergeant would prefer the path going downwards in the military hierarchy,
although there can be other regular paths to its units, e.g., through a lieutenant (see
Fig. 7.6).

What is the situation in other networks? To answer that, let us plot the percentage
of regular paths containing no more than a given number of upstream steps before
going downwards in the hierarchy. In Fig. 7.7 we can compare the results for the
empirical paths to some randomly chosen ones from all the possible regular paths
of the same length. We can observe that the empirical paths contain fewer upstream
steps, which means that those paths try to avoid stepping upwards in the hierarchy.
We can see that around 50% of the empirical paths contain no more than one
upstream step, while the random path's percentage is below 10%. This finding adds
"prefer downstream" as a third identifiable rule that nature seems to consider when
picking a path. So it is not only our military sergeant who should use the downstream
path to issue a command, but this rule seems to be universal and present in other
real-life systems. This finding may sound somewhat contradictory to regularity,
which says that paths should first go upstream, i.e., towards the core, and then
downstream, towards the periphery of the network. However, this is just an apparent
contradiction. The prefer downstream rule only says to pick the downstream path if
available. For example, the bottom part of Fig. 7.5 shows two paths between nodes
X and Y, one beginning with an upstream step followed by several downstream
ones, and one containing only downstream steps, marked as orange and green
paths respectively. In this case, the sergeant can choose between upstream and
downstream regular paths. The prefer downstream rule means that in such cases,
the downstream path is favorable, avoiding stepping upwards in the hierarchy.

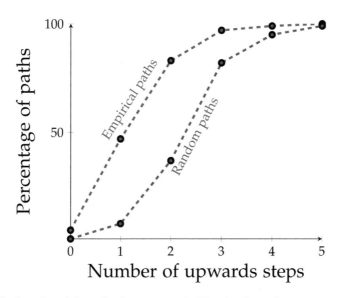

Fig. 7.7 Confirmation of the prefer downstream rule. The plot shows the percentage of regular paths containing no more than a given number of upstream steps before entering the downstream phase. The empirical paths tend to avoid stepping upwards in the hierarchy, which is reflected by the much lower number of upstream steps, in comparison with the randomly selected regular paths of the same length

7.4   Checkpoint

Let's stop here for a second, take a deep breath and summarize our findings about the structure of paths. First, we have seen that empirical paths can be slightly longer than shortest paths. Although in some particular cases, the stretch can be more than four steps, the paths are only around 10–30% longer on average. This inevitably means that nature prefers the usage of short paths. Secondly, we have seen that real networks seem to have an internal logic or an internal hierarchy, which the empirical paths follow in the majority of cases. This simply means that empirical paths go first upwards in the hierarchy and then downwards and paths cannot contain down-up jumps. Finally, paths avoid stepping upwards in the hierarchy if possible, meaning that if there are both downstream and upstream paths available, then the downstream one should be picked.

Our findings hint at the operation of the "prefer short paths", "prefer regular paths" and "prefer downstream" path selection rules. Are these rules equally important, or is one more important than the other? Are there any clear relative priorities among the identified rules? In what follows we argue that there is a reasonable prioritization among those components, which allows us to set up a toy path selection rule set imitating nature's path picking process. According to Fig. 7.3 the prefer shortest path rule can only have lower priority than the "prefer regular path" and the "prefer downstream" otherwise we would not experience stretch at all. Since "prefer downstream" implies the "prefer regular path" rule, the only reasonable choice is to: prefer regular paths at first, then prefer downstream if there is a downstream path and from the remaining paths, prefer the short paths. Remarkably, the length of the paths is just the third thing on this checklist.

7.5   Imitating Nature's Path Picking Procedure

Now let's check how close our argument about path selection above comes to real empirical paths. In order to do this, we define our toy path selection rule set more precisely and compare the selected paths to shortest and empirical ones. We define our toy path selection rule set to:

- **Rule 1** Use regular paths only
- **Rule 2** Pick the downstream paths if available
- **Rule 3** From the paths remaining after **Rule 1** and **Rule 2**, pick the shortest ones
- **Rule 4** If there are still multiple paths remaining break ties randomly

It turns out that the above simple path selection rule set gives very realistic path stretch, close to the stretch computed for the real empirical paths. However, since we explicitly prohibit the use of non-regular paths, all of them are regular, unlike real empirical paths. We have done a hell of a job and made our paths always regular. So, what our simple path selection method cannot reproduce is that empirical paths sometimes violate the "prefer regular path" and the "prefer downstream" rules, although in only a minority of cases. However, the slight randomization of the centrality values of the nodes fixes this. Why would someone randomize the centrality values of the nodes? Well, we have seen that in the case of large and dynamic networks (like the Internet or a social network) we cannot even reconstruct an up-to-date map. Therefore, it doesn't sound reasonable to suppose that any real person or entity representing a node in the network would know the exact structure, i.e., the correct hierarchy of such networks. So, the randomization can be interpreted as simulating the case in which nodes have an approximate picture of the network and therefore their position in the hierarchy is known only with some random error. In that case, the nodes can only have an approximate picture of the internal logic or the hierarchy of the network. This small modification recovers both the stretch and the level of regularity exhibited by the empirical paths.

Our findings are suitable for gaining a vivid estimation of the traffic situation in a large city during busy hours. Monday mornings are always a great stress for the road network and the public transportation system, as everybody goes to work roughly at the same time. On the one hand, due to stretch, empirical paths impose larger average load on the network nodes, on the other, the load is even more concentrated in the inner parts of the network hierarchy (i.e., on nodes located more centrally in the network). The above presented path selection rule set seems to explain the load footprint of real paths better, than resting simply on the assumption that people always using shortest paths. In the context of the public transportation example, the toy policy allows us to better estimate the mass of people at various stations and the possible length of lines at the ticket offices compared to the approximations based on pure shortest paths.

7.6 Two Short Illuminating Stories About Paths

In short, our examinations so far hint at the fact that empirical paths follow the underlying hierarchy (or logic) of the network. They avoid stepping upwards in the hierarchy and they are short, although not always the shortest. How can we make sure that our data cannot be explained by some other path selection rules completely different from what we have found? In short, we can't and this will give us a wealth of possibilities for future research. But we can summarize here two interesting stories investigating paths from a completely different angle, yet come to a remarkably similar conclusion.

An Attracting Story About a Magnet

Anthony was a diligent man. He worked at the subsidiary of a large international relocation company. His bosses took notice of his professional calling early in his carrier, so he went steadily up the ladder becoming a regional leader in his younger years. He was proud to be a fairly autonomous person solving emerging problems on his own.

One day, a peculiar problem arose in connection with the relocation of a whole medical laboratory with some expensive medical equipment. The core of the problem revolved around a specific part of an MRI machine used to monitor the physiological processes inside the human body. Or more specifically, one of its components: a high power magnet that quickly had to be moved overseas, leaving the only possible choice but to carry it by plane. But air cargo companies were reluctant to ship the magnet without a certification issued by an expert stating that transportation by air was safe. Anthony decided to resolve the problem himself; he knew that none of his subordinates had any experience in such matters. However, he knew that Mark, who he did not know personally and was the leader of the Asian branch office, had already been involved in relocating such medical imaging appliances. So he asked one of his friends, Charles, a truck driver at the Asian sub-office to request for help from his boss. In the following week, Charles tried to meet with Mark, but his efforts were in vain; his boss was too busy to make time for him. Eventually, Anthony gave up on Charles and tried sending direct e-mails to Mark, which were also to no avail; it generated no reaction. Ultimately, one last option remained for Anthony: to call the Central Office and ask for some official advice from his bosses. The answer to his questions arrived the very same day. He got the contact details of a university department with experimental physicists who have widely recognized competence in assessing the effects of high power magnets on the navigation system of air flights. With their help, Anthony succeeded in arranging the relocation of the laboratory, although was not happy at all about being forced to resort to his superiors.

The above story perfectly reflects how an employee typically navigates around his organization. The cheapest, and most often, the fastest way, is to look for a subordinate to solve a problem. One should know about their subordinate employees and their capabilities. Also, it is the employee's responsibility to constantly listen to his superiors' orders. They are likely to be the best option when it comes to problem-solving. Turning to superiors, however, takes time, and we also implicitly communicate that we cannot cope by ourselves. You are generally expected to minimize the amount of your boss's time you waste. Finding the help of a co-worker may even be a better option, at least when you have an opportunity to ask a favor of them. Indeed, impeding a superior is the most expensive alternative. Charles, the truck driver, could not even manage to see his boss. Had Anthony known any of Mark's bosses directly, he could have avoided disturbing his own superiors at the Central Office. Finally, we also note that the path from Anthony towards Mark through Charles was not a regular way of connecting the two, as Charles did not have any authority over Mark; he created a non-regular "valley" between the two superiors (Fig. 7.8).

A group of scientists, Peter Sheridan Dodds, Duncan J. Watts, and Charles F. Sabel at Columbia University conducted research in a closely related area some 15 years ago. They studied the information flow in organizational networks [8], for example, the communication of people inside business firms. Their focus was on the robustness of paths of information exchange between entities like departments or individual persons forming the nodes of a network under stress caused by envi-

Fig. 7.8 Organizational hierarchy in the story with the magnet with a path containing a "valley" through a cross-hierarchy edge from Anthony to Mark

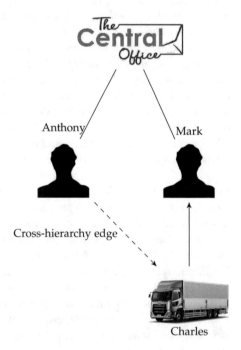

ronmental changes. They were particularly curious about the congestion situations when the network disintegrates according to heavy load on some nodes at strategic points. Just think about a business firm performing poorly due to overloading employees in strategic positions, e.g., overstressed managers. Organizations are supposed to have a strict hierarchical structure according to the relationship of subordination, but they also assume random bonds between individuals, representing informal acquaintances between colleagues. These relationships form additional, so-called cross-hierarchical, edges. When studying this phenomenon, they needed a realistic path selection model, which approximated the load of the nodes in the network to an appropriate extent. After the in-depth study of the literature of organizations, Dodds, Watts, and Sabel settled on a simple **Path selection model** with a three-step decision mechanism at each node traversed by the communication:

Step 1   If an employee of the organization looks for somebody that serves under them somewhere on the working team, then she asks a direct subordinate that connects her to the target person. It may also happen that the direct subordinate *is* the target person, which ends the path right away.

Step 2   If the employee knows a co-worker through a cross-hierarchy edge, e.g., has an informal relationship with somebody who is the superior of the target person or is the target person, then they choose that connection as their second best option.

Step 3   Finally if none of the above holds, then the employee asks their direct superior.

Now, let's examine the structure of paths coming out of the **Path selection model** above. The three-step decision process suggests that if in the organizational network the target person (or target node) is not a subordinate nor an acquaintance of the person holding the message, then they pass the message up in the hierarchy to a direct superior. If the message reaches a node of which the target is a subordinate, the message should be passed downwards in the hierarchy. First upwards, then downwards. Sounds pretty familiar, doesn't it? We have required our paths to go first upwards and then downwards in the hierarchy avoiding containing a large-small-large value pattern in the centrality sequence, thus non-regular paths cannot occur. The **Path selection model** also suggests that if a node is a superior to the target then it should pass the message downwards in the hierarchy, so downstream is preferred in the hierarchy skipping upward steps whenever possible. If Anthony would have applied the **Path selection model** in our story, then he would have solved the MRI relocation problem much quicker, as in Anthony's case the model suggests turning immediately to his superiors.

Let's carry on with our second story about a chaotic elevator system in a multi-story office center.

A Single Story of a Multi-Story Office Center
Our second story is about Kate, a business consultant, who 1 day gets a very interesting job. Her task is to design the layout of offices and working pathways of people in a newly built 50-level office block. The main source of the problem

comes from the fact that a typical work-flow in the company is complicated. The path that should be taken by an employee touches several floors on a daily basis. Additionally, considering the number of offices, there are too few lifts built into the building to serve the requirements. Even though they are large enough to carry a few dozen people simultaneously. It is not the size that counts but the time it takes to get to the caller. If the employees spend half the working hours waiting at the silver doors, the overall efficiency drops to an unacceptable level.

After several weeks of speculation, Kate finally reaches a conclusion. The best solution to cut down delays caused by time-consuming inter-story trips is to design an efficient control algorithm for the lifts. It should be taken into account that the control circuits of the elevators do not have any memory at all. The only information that may be counted on in the movement decision is the direction it was headed right before stopping at a floor: upward or downward. One of Kate's design principles is that unused lifts should always rest near the busiest floors. Furthermore, and most importantly, short waiting times can only be achieved if an elevator, after a stop, always continues to travel in the direction of the nearest floor where the call button is pushed (independent of its previous direction). It doesn't matter much if more people are collected into the same cabin or the elevator takes a few additional detours towards other floors on the way.

After careful planning, Kate has the specially designed lift control mechanism implemented and lets the daily work begin in the office block. Several weeks later she makes a visit to the office to take pride in a job well done. What she observes extremely surprises her. People just gave up taking the elevators and everybody runs up and down the stairs. By questioning a randomly picked employee caught in the stairwell, she learns that the newly designed lift control system generally did prove to be fast enough. Even the amount of energy consumed by the elevators dropped, as the sum of all the paths taken were minimized. However, from time to time people who traveled too far, for example from the top of the building to the first floor, almost never seemed to reach the destination. On the way down there was always a new calling from an upper floor, which was closer than the first floor. On one hand, the inter-story trips were faster on average. On the other, however, people just could not plan the duration of the trip in advance. The system simply became unpredictable. Many times it took only a minute to arrive, but every now and then it lasted more than half a day.

Finally, Kate drew the conclusion that her algorithm failed to fulfill its task and she had to look for some additional advice and redesign the elevator control algorithm.

What can cause such chaos in the office? Kate has a single design parameter in mind, namely the total efficiency of the whole system. She doesn't take into account however, other aspects of the problem. If workers with strict deadlines find a system unpredictable, they rarely venture a trip with an uncertain duration. A similar observation can be taken in many walks of life, especially in the area of engineering: an unpredictable operation of a machinery or an artificial system is rarely beneficial or desired.

In 2001, Lixin Gao and Jennifer Rexford at Princeton University studied the predictability of the Internet routing system [10], which, among computer networking
fellows, is the technical term for the path selection of packets on the network. We
have already seen that paths, which packets take over the Internet, are determined
by routing tables configured in each node by its operating personnel. Since different
nodes may belong to different operating staff working at different networking
companies, they will place their own, independent communication tricks on those
routing tables. For example, to reach a given destination, an Internet company
may prefer to avoid some insecure regions of the Internet. If the companies follow
utterly different rules irrespective of each other, the resulting system may become
chaotic. Some packets may circulate in the network for an indeterminate amount
of time, just like the people in the office elevator in the story above. Due to such
chaotic behaviour, complete regions of the Internet may become unreachable even
if they are connected properly. How can we make the system stable without forcing
individuals to synchronize their actions all the time? Are there any general, but not
too restrictive, rules that the companies on the Internet (or the employees in the
story) should follow, resulting in a tractable system?

Well, Gao and Rexford came to the conclusion that if the Internet companies (the
nodes) agree on some simple and reasonable rules when generating their routing
tables, then the network will behave nicely even if its structure changes in time. The
rules have become famous among networking theoreticians and practitioners under
the name of the "Gao-Rexford conditions". They demonstrated the problem on a
simple object called the dispute wheel [12], a pathological case of chaotic behaviour
widely known among Internet routing theoreticians (see Fig. 7.9).

The solution started by observing that Internet companies act on different levels
of a hierarchy called the "service chain". Do you remember our small Internet
in Chap. 5? Canadian Federal Co. and Maine Trans-Atlantic Co. were at the top
level, and Castle Rock, Salem's Lot and Dunwich were below it. The Gao-Rexford
conditions refer to the acts that should be performed when a packet travels from one
level of the service chain to another.

The Rule of Hierarchy: If a packet comes from an upper level of the service
chain, it should not be sent up again. Or more specifically, the routing tables
should be configured in such a way that packets should never need to turn back
upwards while heading down (see Fig. 7.10).
The Rule of Preferring the Downward Direction: When it is equally appropriate to send a packet up to a higher level of the hierarchy or down to a lower level,
the downward direction should be chosen.

And that is all. If those two simple rules are followed, the Internet is nice and
safe. These two rules are based on a simple observation, that there is an internal
logic, an order on the Internet. A built-in hierarchy of Internet companies. This
hierarchy enables the definition of "up" and "down" and the Gao-Rexford conditions
simply use these directions to prevent packets infinitely circulating in the network.
But there is more. If we meditate on the first rule a bit more, we find that, besides
ensuring system predictability, it has an additional advantage for the Internet as

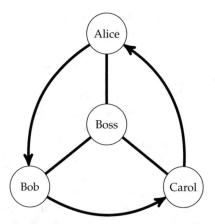

Fig. 7.9 A possible interpretation of the *dispute wheel*, a theoretical object illustrating the unpredictable behavior of communicating actors or nodes making decisions independently of each other not possessing the Gao-Rexford conditions. In the figure, Alice, Bob, and Carol, the employees of an imaginary small organization, communicate with each other, with the intention of passing possibly unpleasant news to their boss. Each of them is reluctant to confront the boss with the bad news, so they all try to persuade each other to relay the message to the boss, but none of them actually does so. The wheel exemplifies that the message never arrives at its destination, even so, the nodes in the network are well connected

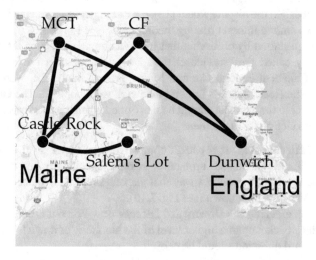

Fig. 7.10 Our previously developed tiny model of the Internet initiated by the people of Castle Rock to communicate with the outside world by connecting to nearby town Salem's Lot and Dunwich in England using the transit services of Main County Trans-Atlantic (MCT) and Canadian Federal Co. (CF) as a backup route

well. We require that packets heading towards companies at lower layers of the hierarchy cannot be sent up again. That means it cannot happen that two higher layer companies communicate through a lower layer. Canadian Federal Co. and Main County Trans-Atlantic, the companies handling high traffic volumes, may never use the, possibly quite lightweight, networking infrastructure of Castle Rock. That sounds quite reasonable that majors should not overload the system of the lesser. They should use a direct connection or even larger transit companies to make the connection.

Applying the Rule of Preferring the Downward Direction also has some additional advantage that we can uncover with a little analysis. In the hierarchy of the Internet, lower layer companies pay the higher ones to be connected to the big network, pretty much like your home Internet subscription. You pay your provider to forward your data to the Internet, but your provider won't pay you for receiving data. Thus, the connection is free for the one being higher in the hierarchy. Which one is the better choice? Using a free connection down to an inferior party or going upstream for a price instead? Let's prefer the downstream!

Okay, now can we apply these magic rules for the elevator problem as well? Let's continue and see how Kate finally solves the problem by studying some advanced computer networking principles. After a more intensive investigation of the business procedures within the company, Kate realizes that the employees' inter-story paths show some orderliness. Most of the time they visit one or more of their superiors then return. So, a better office arrangement would be to move the employees to different levels of the building according to their position in the chain of command. More senior officials should be put on the higher floors, with the general manager at the top. In such a way, the routes of the employees become less random, the movements can be harmonized a bit more. In short, we build a hierarchy in the office according to the journeys of the employees.

Now, it is not too difficult to see the parallels in the two different areas of engineering. By rephrasing the elevator problem in new terms, we are ready to relieve the difficulties of the office block: the elevator makes a decision every time it stops at a floor during operation. Those decisions depend on the demands (i.e., destination floors) of the actual travelers and the current direction of the elevator, independent of previous decisions due to the lack of memory. If you recall how Internet routing works (see Chap. 6.3), you find the situation very similar. Now we can match the control decision of the elevator with the forwarding policy of the Internet companies in sending the packets to lower or higher hierarchy levels. Both decisions are *local* in time, meaning that past decisions along the route cannot be taken into account. It doesn't matter how complicated the lift control mechanism is: if at a minimum, it keeps itself to the Gao-Rexford principles, the resulting global behavior remains predictable.

The Rule of Hierarchy tells us that we should never change the recent direction of the elevator even if one of the current travelers decides so. Unless, of course, there are no further demands to keep going in that direction. In such a way, none of the travelers need to take unnecessary detours, that is taking up-down-up or down-up-down "valleys" during their path as the Rule of Hierarchy forbids the down-up pattern. Of course, an employee may lose their temper while waiting for the elevator and push the call button for both directions announcing an ambiguous request. In such a case, she is doomed to be taken on a possibly longer journey, infuriating her even more.

Now consider what happens if the control mechanism loses the information of the current direction, for example, due to a power failure or a circuit damage. This is when the Rule of Preferring the Downward Direction comes into life. If there was not a default action included in the control mechanism for such cases, the lift might still roam up and down for eternity between neighboring floors changing its mechanical mind each time. The choice for the downward direction for an elevator as a default action seems to be more advantageous yielding an escape route in the course of emergency e.g., fire or earthquake, should the careless employee decide upon using the elevator in such a case.

Had Kate been aware of the above two rules when designing the elevator control system in the office center for the first time, the employees would have been much less fatigued when finally getting home at the end of the day. How fortuitous that elevator companies in the real world generally stick by it!

Now we can see that Gao and Rexford came to remarkably similar path selection rules to ours, albeit their target was to ensure the predictable operation of the Internet via theory. Although the definition of the hierarchy is not the same as ours, the Rule of Hierarchy sounds familiar to our finding that empirical paths rarely contain a large-small-large pattern anywhere in their centrality sequence. Although at this time it is not clear if the Gao-Rexford conditions can be mapped exactly to our centrality-based rules, it is worth wondering whether our observations about the stretch of paths enable the predictability of operation in other networks similar to the Internet. Additionally, also the Rule of Preferring the Downward direction has a ghostly resemblance to our observations on paths in real-world systems.

These short stories have identified similar logic of path selection in two distinct fields of networks. We found something very similar through measurements and not just only on the Internet but also in three networks coming from very different corners of life. Can all these still be a mere coincidence? In these stories, the authors use very different methods and definitions for constructing the underlying hierarchy and they formulate the rules of path selection quite differently. Our arguments about the structure of paths generalize these sporadic results and show that there can be an amazing universality in path selection across many different networks. Path selection rules built upon centrality creates a common ground for speaking about underlying hierarchies. Thus reasoning about networks and paths coming from possibly arbitrary fields of life is possible. Now we have a rough picture about the structure of real-world paths.

The reader at this point can ask the very reasonable question: So what? We have identified some characteristic similarities of path selection in various systems coming from a wide spectrum of life. What can we do with all this? For what purpose can we use it? Well, before delving into this, we suggest sitting back for a second and just simply being amazed by the possible universality of the nature of paths. One should never underestimate the benefits of just sitting back, relaxing

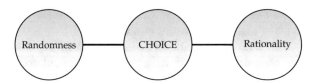

Fig. 7.11 Paths in nature lie between pure randomness and pure rationality

and being amazed by something. So, paths in real life are clearly not the results of pure randomness. They are not like mucking about in the city. Quite the contrary, they are short and right on target. On the other hand, paths are not governed by pure rationality either. They are not always the most efficient; those would be the "soulless", monotone and dry shortest paths. They are somewhat stretched. Real paths exist somewhere in the middle of these two extremes.[2] Now, what is the similarity between pure randomness and pure rationality? There is no possibility of choice in either of them. When our paths are governed by randomness, we are "just floatin' around accidental-like on a breeze",[3] while at the other extreme, pure rationality dictates taking the path with the highest supposed benefits or lowest cost, depriving us of our free will. So, it seems our paths lie in between. Although our results suggest they are closer to rationality. Recall, that our empirical paths were only around 10–30% longer than the shortest paths, so real paths are nearly as effective, or energy-efficient as shortest paths. But while being efficient and target oriented, real systems can still make a choice. The 10–30% enables us to form messages about the way we live our lives, our way of solving problems. Through these little extra steps, we are allowed to make a difference and form our own stories. 10–30%. This is our playground.

Even ancient Chinese wisdom has something interesting in store for us about choosing a path of appropriate length. The more than 6000-year-old art of Feng Shui is about harmonizing people with their direct or indirect environment. It is actually one of the many Taos of the big Tao, just like the art of tea ceremonies or paysage. According to Feng Shui, everything in the world, life or lifeless, is a form of liquid energy, the Chi, that also flows constantly around any objects, connecting them together. It streams inside and outside of things, nothing can block its way, not even the thickest concrete wall. Objects, however, can modify its direction or speed, which can alter the quality of energy. By rearranging the objects that surround us, creating appropriate *paths*, we can avoid negative energy and increase positive energy, which can have a beneficial effect on our mood, emotions, or even our

---

[2]See Fig. 7.11.

[3] I don't know if we each have a destiny, or if we're all just floatin' around accidental-like on a breeze But I, I think maybe it's both.

— Forrest Gump.

health. The course that the ever-flowing Chi takes has a crucial influence on the energy. So, what is it, that Feng Shui tells us about paths?

At this part of the book, you should not be surprised that Feng Shui is absolutely against shortcuts. Straight pathways or trails should not be used to connect objects, for example, in your garden. According to the Tao, it is because the flow of Chi becomes speedy, streaming too fast out of your garden. But trail crossings or sharply curving paths just like sharp corners or other saliences have a negative effect, sending forth "poisonous arrows" (negative energy or Sha) blocking the flow of Chi. Straight short paths are not correct, but lengthy curving driveways are equally wrong. Brightly winding watercourses are of less positive effect, but cutting down the curves also emits Sha. The ultimate advice of Feng Shui is to form slightly curved pathways in your garden (Fig. 7.12). Similarly to the curved middle path in the Yin Yang symbol (Fig. 7.13), if it finds you in an even more philosophical mood.

Fig. 7.12 A curved pathway in the Japanese garden of the Budapest Zoo overlayed with an artificial pathway constructed by joining two segments each being a third of a circle. Walking along the synthetic path makes the distance between the two endpoints around 20% longer. The photo is the property of the authors

Fig. 7.13 The Yin Yang, a
Tai Chi symbol with the
indication of the middle path
by a red line

Chapter 8
Amazing Scientific Discoveries: Aspirin, Cattle, Business Communication and Others

Besides the philosophical arguments, one could think of a series of possibilities where all these observations about paths could be used. Have you ever wondered, for example, what happens in your body after you swallow an Aspirin? As incredible as it sounds, this question was answered only 74 years after the development of the medicine. It was 1897 when the young German chemist Dr. Felix Hoffmann managed to stabilize the agent of Aspirin. After patenting in Germany in 1899 and in the US in 1900, Aspirin started its great triumph and became the most popular painkiller worldwide. Even Neil Armstrong took an Aspirin pill in his medical-kit when going to the Moon on the Apollo 11. In the early 1970s, more and more researchers asked the question: How and where does Aspirin work in the body? Pharmacologist Sir John Vane was the first to demonstrate the classical effect profile of Aspirin, for which he received the Nobel Prize in 1982.

What is this foggy mystery about the effect of drugs? Well, the effects and side-effects of drugs are mainly characterized by the path of molecules which the drug interacts until it has its targeted effect. Usually, the drugs are first converted into so-called metabolites which then interact with the metabolic network of the cells.[1] The metabolic network is comprised of the maze of chemical reactions in our cells over which various materials are converted into each other. A specific path or set of paths in this network can basically correspond to a chain of chemical reactions happening after somebody swallows a pill. For example, after taking an Aspirin, it is readily hydrolyzed to salicylic acid, which in turn undergoes conjugation reactions generating the major metabolites salicyluric acid and glucuronides. And this is just the major path of Aspirin's metabolism, there are other minor paths governing the whole metabolic process and thus the effects and side effects of Aspirin. Interestingly, simple shortest paths in the metabolic network do not always reflect the biochemical facts. Such paths may introduce biologically infeasible

---

[1] See, Fig. 8.1.

© The Author(s) 2021
A. Gulyás et al., *Paths*, https://doi.org/10.1007/978-3-030-47545-1_8

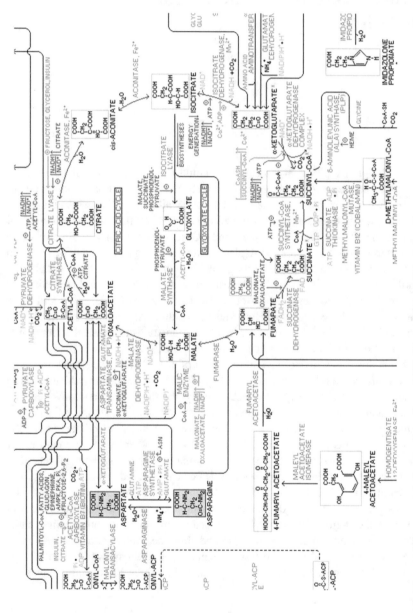

Fig. 8.1 A small part of human metabolism by Evans Love. [With the permission of Evans Love]

shortcuts [18]. Thus, a deeper understanding of the structure of paths can be used to estimate the side effects of drugs more carefully even before anyone takes them.

The sum of everyday paths taken by people in larger cities to reach their workplaces or homes constitutes the load which public transportation and road systems have to carry day-by-day. The appropriate knowledge about these paths can support the design and operation of such systems and approximate their behavior in unforeseen situations, such as scheduled network changes, roadworks, natural disasters or walkouts. In the era of in-pocket GPS powered route planners, the assumption that people use the shortest paths for traveling between their sources and destinations seems more than reasonable, it seems somewhat obvious. Recently, Shanjiang Zhu and David Levinson at the University of Minnesota decided to check the validity of this assumption [27]. They evaluated the paths followed by residents of the Minneapolis-St. Paul metropolitan area and they had an interesting observation. For some reason, people don't always use the shortest possible path for their journeys. They found that, if the destination is near (around 1.5 km), 80% of people follow the shortest paths, the remaining 20% prefer a longer ride. Interestingly, if the destination lies farther, the larger portion of people tend to take a longer ride compared to the shortest possible path. For example, if the destination is around 16 km away, then only around 17–18% of people follow the shortest path and the majority of them will choose a longer path. Thus, the authors' claim is that the available path selection methods based on the shortest path assumption cannot reveal the majority of paths that individuals take and they promote future efforts for building better path selection models and improving transportation services based on them.

It seems that it is not just humans who like detours. Temple Grandin, the famous autistic scientist, discovered that cattle like to go on curved routes and they are reluctant to cross straight passages[11] for some reason. Grandin's ability of thinking in pictures enabled her to discover the main motives and fears of cattle when put into a pen system. Her designs of pen systems have revolutionized the industry and today her patterns are widely adopted in cattle-handling facilities across the US. In such facilities, an important procedure is dipping the cattle in a vat filled with water to free them from ticks. The common way of doing this was to order the cattle to the vat through a straight corridor made of cattle-pens. Despite their good swimming skills, many cattle drowned in the course of dipping, so Grandin redesigned the whole pen system to make it more comfortable for the cattle (see Fig. 8.2). She observed that cattle like to move along curved paths, similarly to Chi. So instead of directing the cattle to the dip vat through a straight path, she used slightly longer and curved races. These curved segments have calmed down the cattle, significantly decreased their level of stress, thus decreasing the possibility of drowning. As of today, it is estimated that half of the cattle in the United States and Canada are handled with equipment Grandin has designed.

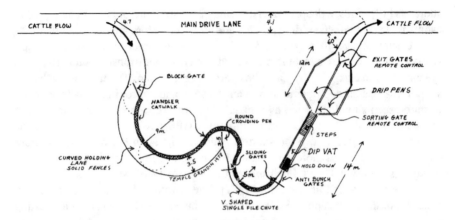

Fig. 8.2 Handling system for dipping cattle with curved races. As appeared in the publication [11]. [With the permission of Temple Grandin]

Similarly to directing cattle in pen systems, the flow of information among employees could be directed within a business firm. Why not? A prerequisite of good problem-solving in a business organization is good communication between the employees. We have seen in Chap. 7.6 that an unnoticed dispute wheel (which is nothing more than a bunch of employees/decision-making entities using each other as strawmen, communicating in a circle and passing the information infinitely among each other) can lead to a communication disaster. The structure of an organization could surely be enhanced based on information about the communication paths preferred by people. For example, business firms can employ professional communication strawmen, whose task it is to lower communication boundaries between people and ensure a more fluent and less stressful communication culture within the company. Just think about public relations (PR) specialists, whose job it is to maintain positive relationships between the company and the public. As we have seen in Chap. 7.6, organizational networks usually have a strong backbone hierarchy on top of which the informal links between employees appear [3]. In such networks, the paths of communication and problem-solving coincide with our findings about paths in other networks. Although social networks seem much more complicated, there is a chance that we will find similar characteristics in the paths of retweets on Twitter[2] or in the sharing paths of posts at Facebook. Such knowledge about paths in social networks can be used to strengthen and better organize the human communities. The way the human brain processes information and learns is one of the great mysteries of life. In lack of path related data from inside the brain, neuroscientists mostly use shortest paths when reasoning on information processing paths in the brain. However, with the adoption of modern computer-aided imaging and measurement techniques (like DSI and fMRI), more realistic models of paths

---

[2]see Fig. 8.3.

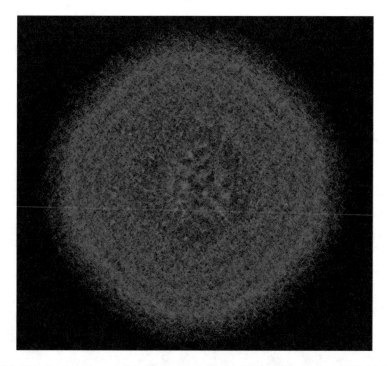

Fig. 8.3 Visualization of a part of Twitter by Elijah Meeks. [With the permission of Elijah Meeks]

selection are proposed. For example, in a very recent work, Koenigsberger and others argue that brain paths should be somewhere in between random diffusion and shortest paths [1]. This observation highly coincides with our results in other networks. Although the end of this path seems very far, the proper characterization of neural paths inside the brain can take us a step closer to solving the challenging mystery of the human brain.

Chapter 9
Paths to the Way We Live, Teach and Learn

In addition to the scientific applications, are there any personal benefits for the reader? Are there any messages which can be kept in mind that would affect our everyday lives? In order to answer this, consider Fig. 9.1. This figure represents an abstract map of a city showing its core at the top and boundaries at the bottom and also some strategic public transportation points where one can switch between transportation devices (buses, trams, underground, etc.). Now let's assume that we are newcomers to the city without any knowledge about the possible transportation options. Our first trip on our first day in the city is to travel from our Hotel (H) to the Bank (B). Now let's assume that we have a "path oracle" (e.g., tourist information) which can be asked for paths between arbitrary points in the city. This oracle always recommends the shortest path between the source and destination. For the trip between the Hotel and the Bank, the oracle gives us the green path, which requires three changes at intermediate transfer points and is the shortest in terms of distance. So, from now on, we can reach the Bank from the Hotel and withdraw money. On the next morning, we want to see the Castle (C), for which the oracle gives us the olive colored path. This has similar characteristics to the green path: three changes and a small traveling distance. On the third day, we plan to withdraw some cash at the Bank and than head straight to the Castle, but when we ask the tourist information, we realize that they are closed. So we are left on our own. What can we do? Well, we know the shortest path from the Bank to the Hotel, and another shortest path from the Hotel to the Castle, so we can travel from the Bank to the Hotel on the green path, then from the Hotel to the Castle on the olive path. This journey will take seven changes at various transfer points.

Now let's turn back the hands of time and assume that we are again newcomers and do the same thing but with a slightly different oracle. This second, more professional, tourist center gives us short paths that preferably follow the hierarchy of the city, i.e., first heading towards its center (upwards), then heads sideways to the border of the city. For our first day trip between our Hotel and the Bank, this second oracle reveals the magenta path trough the Town Hall (T). This trip requires three changes, but slightly longer traveling distance compared to the shortest path

© The Author(s) 2021
A. Gulyás et al., *Paths*, https://doi.org/10.1007/978-3-030-47545-1_9

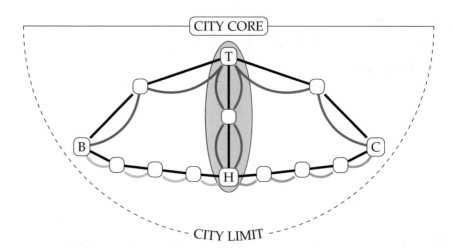

Fig. 9.1 Role of shortest and regular paths

as this goes into the city first and then out to the destination. On the second day, the oracle tells us the red path for our trip between the Hotel and the Castle, which has pretty similar characteristics to the magenta path, also passing by the Town Hall. In the morning of the third day, we need to go from the Bank to the Castle but the employees of the tourist center went on a team-building activity. Again, we are alone in finding a way around! Now if we combine our two paths from the Bank to the Hotel and from the Hotel to the Castle we obtain the following route plan: Bank $\to \ldots \to$ Town Hall $\to \ldots \to$ Hotel $\to \ldots \to$ Town Hall $\to \ldots \to$ Castle. But wait a minute, we made an unnecessary by-pass between the Town Hall and the Hotel. We can completely skip this loop and finally take our trip by using the Bank $\to \ldots \to$ Town Hall $\to \ldots \to$ Castle. This path requires three changes and only slightly longer traveling than the shortest path. So, relying on the advices of the second more professional tourist center in the first 2 days, we can come up with a pretty nice and short trip on the third day all by ourselves.

How can it be? Well, we can see that in the second case, when the oracle gives us paths coinciding with the underlying logic of the city, we can learn much more about the city compared to the case when we stick strictly to the shortest paths. The overlap of the two paths[1] gives the possibility to relate the paths to each other and from these relations, we extract greater knowledge. After a few initial trips following the logic of the city, tourists can usually leave their city maps in the hotel because they can travel between almost any sources and destinations on the map. Now, imagine how you guide somebody who asks for a path towards some place in your city. Do you offer them the shortest path? Well, maybe you don't know the shortest path to the destination after years of living there. But most probably you will offer a path

---

[1] Marked with a pink ellipse in Fig. 9.1.

coinciding with the logic of the city, i.e., transfer somewhere downtown. Isn't it amazing that nature seems to implement exactly this in many real-life systems? The Internet, the air transportation network, even the human brain seems to implement paths like these. Nature is wise enough not to use shortest paths to ensure knowledge accumulation and navigation inside the system. Long paths cannot be used as they are not effective. So, nature implements a trade-off. By using 10–30% longer paths, it ensures integrity, learnability, and navigation.

Now let's put this imaginative story into a broader context. Consider that we would like to quickly train somebody to travel inside the city by using the public transportation services. In fact, the knowledge we would like to give her is a usable network of transferring points and their relations to arbitrary points in the city. We can take her with us on some trips within the city. What trips should we choose? Well, we would be surprised if somebody would use shortest path trips for this purpose. What is more probable is that we choose trips according to the logic of the city and in a fashion that the back-to-back journeys have common points preferably in the downtown area. These paths obey the hierarchy or the internal logic of the city and will maximize the extractable knowledge over the shortest period of time.

Now suppose that we want to teach something arbitrary (e.g., biology, maths or farming) by giving lectures or presentations. We argue that the knowledge we want to share is not fundamentally different from the knowledge of being a good user of public transportation. In fact, there are studies showing that children with stronger navigation skills among the world's physical objects can perform better in various learning tasks [19, 23, 26]. Although instead of the physical nodes and physical connections (changing stations and various connecting lines), our nodes and links will likely be "virtual", for example notions and relations between them. The knowledge that we would like the audience to have is a network of concepts related to the topic; we would like them to know the main notions of the field and their connections to each other so they can navigate easily between them and come up with their own ideas as soon as possible. What kind of sentences should we choose? How should we structure our teaching materials?

Emma Ledden, author of the #1 selling "The Presentation Book"[16] says that: "To avoid your presentation becoming muddled, structure it around your core message or messages. These should be illustrated in different ways, revisited and emphasized, so they are understood and remembered." So, during your talks or lectures, you should frequently revisit the core messages and illustrate them in different ways. This sounds quite the same as showing our imaginary friend a variety of trips in the city, with common points in the city center, isn't it? So nice presentations introduce new elements by carefully relating them to the central notions the presentation is built around and thus revisiting the core concepts many times. These "mental" paths over virtual nodes are the exact equivalent of our city trips in the physical world joining the source and destination nodes through strategic intermediate nodes in the city center.

Now we can turn back for a moment to our question raised in the introduction about mind map based presentation tools: what allowed these tools to compete with giants of the IT industry in the area of presentation software? Well, these "revisiting" mental paths are nicely visualized by the mind map based tools when zooming from one view to another, which amplifies the memorization of these paths. Showing the context graphically, the audience is visually taken on a path from notion to notion through the whole logical structure of the presentation, similarly to the trips in our imaginary city. We argue that this path visualization feature of mind map based presentation tools can account for their increasing popularity among presenters.

How can we learn something effectively? If we can use paths to increase the efficiency of teaching, it sounds pretty reasonable that we can use them to increase the efficiency of learning too. How can we use paths to learn something? Well, if we accept the idea that the lecturer is trying to communicate a network, through some specific paths they picked, we can make our learning more efficient by enhancing the process of building this network in our mind. For example, when listening, we can continuously try to relate the things we hear to each other. Moreover, we can create quick experiments to see if we can relate some randomly picked concepts on our own. These experiments can enable us to learn how to navigate among the concepts of the field and discover new connections independently. The formulation of good questions can also be aided by such experiments. In our understanding, learning (not lexical, but practical) is pretty much the same as navigation. Do you want to be better at learning? Our findings about paths say: turn off your GPS whenever you can! Forget about always using the shortest paths to reach your goals! You want your children to learn better at school? Our observations say: Let them tramp freely after school (preferably in a safe environment) and self-develop their navigation skills.

There is a famous Hungarian saying: Don't leave the betrodden way for the untrodden one! It is the subject of continuous misinterpretation in Hungary. People usually think that this saying suggests that you should always be on the safe side and you shouldn't try out anything new. This is usually considered as the curse of Hungarians, according to many thinkers, as it seems to provide the recipe for an unsuccessful life. Delving deeper into the real wisdom this saying may offer us something about learning things more efficiently in the long term. Obviously, when you are in a hurry, you attempt to use the shortest path from your current position. Such a path, however, is likely to be an untrodden one. Think about, for example, finding your way through a big city park from its entrance towards one of its exits. It is not likely that a paved road connects the two in a straight line: you have to tread on the grass. The more you walk on the grass, the higher the possibility that you miss a few signposts that stand only by the paved roads. On one hand, you may miss your direction, on the other, even if you find your way, you would not learn anything about the whereabouts of other exits. Surely if you cross the park only once, late for your own wedding, you would not really care. However, if you do that on a daily basis, you will hardly ever learn what the park really looks like. By skipping the overlaps between your paths, you will never see the relation between your paths and you will hardly learn to navigate yourself confidently. If you pick the betrodden instead, i.e., the sometimes curvy, paved roads in the park, you increase the possibility of overlapping and the chance to learn to orient yourself. Preferring the betrodden doesn't mean by default that your paths should not contain untrodden parts, which is quite inevitable in life. You continuously discover new places and people, thus it is highly likely that you will have to use at least partially unknown paths to reach them. But the wisdom of this saying is to keep to the betrodden path while it is possible, to relate and integrate the untrodden segments properly, and to finally understand what the big picture really looks like.

Chapter 10
The Path is the Goal!

Our one sentence conclusion about the nature of paths was that they follow an internal logic of the underlying network even if this comes at the price of being slightly longer. Does it imply that our paths will be the same? Does it imply that our behavior will be deterministic and totally constrained by the network? Before contemplating these questions let's think a bit about the scenario when there is *no* network.

If you have ever walked through a public park you may have noticed that besides paved ways there are many unpaved path segments which are clearly used by people according to the trampled grass (despite the "Keep off the grass!" warnings). In fact, modern parks are paved only after a few months of public usage and the paving follows the trampled paths of people. In this case, there is no network which dictates the logic for the paths. But after a few months of public usage, a clear network is formed. What happens here? Well, people start using the park in their own way. None of them use it in the same way since they usually enter and exit at different points of the park, thus their behaviour inside the park is different. They will have varying preferences about the things in the park. Some people are interested in the statues, others seek benches under shady trees or the public workout areas. The network which is finally paved emerges from the summation of people's interactions with the park. But this doesn't mean that they will start to use the same paths after they are paved. The network of paved segments in the park can be easily reconstructed after a few hours or days of walking depending on the size of the park. In fact, such maps are usually placed at the entrances showing the main attractions and roads inside the park.[1] This map acts as a kind of public information. What about the paths? Well, the paths still belong to the people. The paths describe the habits of the people and tell us about them. About their favourite places, the location of their home and even about their health, if they prefer long or short walks. They still use the park in their own special way.

---

[1] See Fig. 10.1.

A. Gulyás et al., *Paths*, https://doi.org/10.1007/978-3-030-47545-1_10

Fig. 10.1 The official map of
Central Park in New York
City. [With the permission of
the Central Park
Conservancy]

Similarly to your specific footprints in a public park, your paths seem to define you in a more general way. We have seen that there is not too much choice in shortest paths while long paths may be boring and unfollowable. The game of expressing yourself lies somewhere in between. How do you communicate? Where do you choose to transfer? What kinds of lovely and memorable detours have you taken in your life? These small detours seem to define you. Your paths are you. People can be identified by a short sequence of consecutive web pages they visit or by simply observing the movement path of the mouse. Even without their fingerprints or retina scans. The way people choose their letter changing words in the word-morph game seems to be unique and specific for each person. That is the power of paths. It is understandable that most studies considering paths in various systems simply suppose that shortest paths are used. Well, that is the simplest, clearest and most rational way of thinking. Even if they detect detoured paths by measurements, they usually argue that there is something missing in the network model. There should exist some twist in the network (e.g., adding weights to the edges which will stretch or shorten them), which will explain the real-life paths as shortest paths in a modified system and the world will be understandable again. But this way of thinking eliminates the possibility of choice from path selection and sterilizes the problem by eradicating all forms of life and spice from the scenery. **Our take-home message in this book is that detours are not bugs, they are a feature**. A detour is like a signature. A detour is something that adds the spice, which adds the story, which adds the logic, which adds the choice, which adds vitality, which adds the meaning to the sequence of paths. What are the most important strategic core points of your life against which you relate all your events? What is the internal logic or the underlying hierarchy of your life? What are your core messages? These can be many things, your family, your friends, your hobbies, your career, money, power even alcohol or drugs.

It seems that we cannot change our starting point in life and neither its end. As Eric Berne, the famous psychiatrist sarcastically said: "life is mainly a process of filling in time until the arrival of death, or Santa Claus, with very little choice, if any, of what kind of business one is going to transact during the long wait". But still, these little choices tell us who we are. By choosing these intermediate points and taking our small but specific detours, we are allowed to make a difference. Our path is more important than our destination. *The path is the goal.*

Coda

One day one of the monks came unexpectedly into Linji's cell.[1]

"Master!" He exclaimed excitedly. "I spent a year in your monastery studying the Doctrine. I got to know the ideas of the Three Carriages of the Teachings, the Regions of the Triple World, the Creatures of the Sixth Worlds of Existence, I went through the ten states and the realms of the thousand things, I am over the Theorem of Four Corners and beyond the hundred denials but I still cannot answer the question: Who am I? Answer me, Master, who am I?"

Linji looked up at the monk then turned away from him. So he asked, after a long pause, softly.

"Did you eat your porridge, monk?"

"I ate it, Master."

"Have you washed your dining-tray?"

"I washed it."

"Did you clean your cell?"

"I did, Master."

"Did you sweep the yard?"

"I swept it, Master."

"Did the leaves come when you swirled, huh?"

"They did, Master."

"But you got them together."

"I got them, Master."

"And did you see the big cypress tree in the yard?"

"I saw it, Master."

"It's a nice tower-like, right?"

"Yes it is, Master."

"It has a delicate, balmy scent. Did you smell it?"

"I smelt it, Master."

---

[1]A story translated into English by the authors from the book: Su-la-ce. *Reggeli beszélgetések Lin-csi apát kolostorában, közreadja Sári László* Kelet, (2013).

© The Author(s) 2021
A. Gulyás et al., *Paths*, https://doi.org/10.1007/978-3-030-47545-1

"Then you came to me."

"I came to you, Master."

"And you asked who you are?"

"I asked, Master, who am I."

"Well, I want you to know, monk, you're the one who ate his porridge in the morning, washed his dining tray, cleaned his cell, swept the yard, watched the leaves go and smelled the balmy scent of the tower-like cypresses. Then he came to me and asked who he is. Well, this is you, monk, nothing else. But this is enough, believe me. A man cannot be more than that." Linji said, turning to the monk and looking him slowly in the eye. Then they just stood there silently for a while. The monk felt that the Master was honest with him. He bowed long before him and returned quietly to his cell.

Bibliography

1. Andrea Avena-Koenigsberger et al. "A spectrum of routing strategies for brain networks". In: *arXiv preprint arXiv:1803.08541* (2018)
2. Albert-Laszlo Barabasi. *Linked: How everything is connected to everything else and what it means for business, science, and everyday life*. Plume, 2003
3. Tom E Burns and George Macpherson Stalker. "The management of innovation". In: (1961)
4. Center for Applied Internet Data Analysis. *Internet Traceroute Database*. www.caida.org
5. Thomas Frederick Crane. *Italian popular tales*. Boston, New York: Houghton, Mifflin, and Company, 1885, pp. 252–253
6. Attila Csoma et al. "Routes Obey Hierarchy in Complex Networks". In: *Scientific Reports* 7.1 (2017), p. 7243
7. E. W. Dijkstra. "A Note on Two Problems in Connexion with Graphs". In: *NUMERISCHE MATHEMATIK* 1.1 (1959), pp. 269–271
8. Peter Sheridan Dodds, Duncan JWatts, and Charles F Sabel. "Information exchange and the robustness of organizational networks". In: *Proceedings of the National Academy of Sciences* 100.21 (2003), pp. 12516–12521
9. S N Dorogovtsev and J F F Mendes. *Evolution of Networks: From Biological Nets to the Internet and WWW*. Oxford: Oxford University Press, 2003
10. Lixin Gao and Jennifer Rexford. "Stable Internet routing without global coordination". In: *IEEE/ACM Transactions on Networking (TON)* 9.6 (2001), pp. 681–692
11. Temple Grandin. "Observations of cattle behavior applied to the design of cattle-handling facilities". In: *Applied Animal Ethology* 6.1 (1980), pp. 19–31
12. Timothy G Griffin, F Bruce Shepherd, and GordonWilfong. "The stable paths problem and interdomain routing". In: *IEEE/ACM Transactions on Networking (ToN)* 10.2 (2002), pp. 232–243
13. John Guare. *Six degrees of separation: A play*. Vintage, 1990
14. Frigyes Karinthy. "Chain-links". In: *Everything is different* (1929)
15. Attila Kőrösi et al. "A dataset on human navigation strategies in foreign networked systems". In: *Scientific data* 5 (2018), p. 180037
16. Emma Ledden. *The Presentation Book: How to create it, shape it and deliver it! Improve your presentation skills now*. Pearson Education, 2014
17. Jure Leskovec and Andrej Krevl. *SNAP Datasets: Stanford Large Network Dataset Collection*. http://snap.stanford.edu/data. June 2014
18. Hongwu Ma and An-Ping Zeng. "Reconstruction of metabolic networks from genome data and analysis of their global structure for various organisms". In: *Bioinformatics* 19.2 (2003), pp. 270–277

19. Wenke Möhring, Andrea Frick, and Nora S Newcombe. "Spatial scaling, proportional thinking, and numerical understanding in 5-to 7-year-old children". In: *Cognitive Development* 45 (2018), pp. 57–67

20. Seiji Ogawa and Tso-Ming Lee. "Magnetic resonance imaging of blood vessels at high fields: in vivo and in vitro measurements and image simulation". In: *Magnetic resonance in medicine* 16.1 (1990), pp. 9–18

21. OpenFlights. *Airport Database*. www.openflights.org

22. Vivek S Pai et al. "The dark side of the Web: an open proxy's view". In: *ACM SIGCOMM Computer Communication Review* 34.1 (2004), pp. 57–62

23. Arnaud Saj and Koviljka Barisnikov. "Influence of spatial perception abilities on reading in school-age children". In: *International Journal of Psychology* 51 (2016), p. 78

24. Péter Szabó. *Don Bend*. Zrínyi Publishing Group, 2013

25. Duncan J Watts. *Small worlds: the dynamics of networks between order and randomness*. Princeton university press, 1999

26. Ahmad Yarmohammadian. "The relationship between spatial awareness and mathematic disorders in elementary school students with learning mathematic disorder". In: *Psychology and Behavioral Sciences* (2014)

27. Shanjiang Zhu and David Levinson. "Do people use the shortest path? An empirical test of Wardrop's first principle". In: *PloS one* 10.8 (2015)

Printed in the United States
by Baker & Taylor Publisher Services